OXFORD MASTER SERIES IN CONDENSED MATTER PHYSICS

OXFORD MASTER SERIES IN CONDENSED MATTER PHYSICS

The Oxford Master Series in Condensed Matter Physics is designed for final year undergraduate and beginning graduate students in physics and related disciplines. It has been driven by a perceived gap in the literature today. While basic undergraduate condensed matter physics texts often show little or no connection with the huge explosion of research in condensed matter physics over the last two decades, more advanced and specialized texts tend to be rather daunting for students. In this series, all topics and their consequences are treated at a simple level, while pointers to recent developments are provided at various stages. The emphasis is on clear physical principles of symmetry, quantum mechanics, and electromagnetism which underlie the whole field. At the same time, the subjects are related to real measurements and to the experimental techniques and devices currently used by physicists in academe and industry.

Books in this series are written as course books, and include ample tutorial material, examples, illustrations, revision points, and problem sets. They can likewise be used as preparation for students starting a doctorate in condensed matter physics and related fields (e.g. in the fields of semiconductor devices, opto-electronic devices, or magnetic materials), or for recent graduates starting research in one of these fields in industry.

Soft Condensed Matter

R.A.L. JONES

Department of Physics and Astronomy
University of Sheffield

OXFORD

UNIVERSITY PRESS

OXFORD
UNIVERSITY PRESS

Great Clarendon Street, Oxford OX2 6DP

Oxford University Press is a department of the University of Oxford.
It furthers the University's objective of excellence in research,
scholarship, and education by publishing worldwide in

Oxford New York

Auckland Cape Town Dar es Salaam Hong Kong Karachi
Kuala Lumpur Madrid Melbourne Mexico City Nairobi
New Delhi Shanghai Taipei Toronto

With offices in

Argentina Austria Brazil Chile Czech Republic France Greece
Guatemala Hungary Italy Japan Poland Portugal Singapore
South Korea Switzerland Thailand Turkey Ukraine Vietnam

Oxford is a registered trade mark of Oxford University Press
in the UK and in certain other countries

Published in the United States
by Oxford University Press Inc., New York

First published 2002
Reprinted 2003 (with corrections), 2004, 2006 (twice), 2007, 2008, 2009 (twice),
2011 (twice), 2012, 2013 (twice)

A catalogue record for this title
is available from the British Library

Library of Congress Cataloging in Publication Data
(Data available)

ISBN 978 0 19 850589 1 (Pbk)
ISBN 978 0 19 850590 7 (Hbk)

14

Typeset by Cepha Imaging Pvt. Ltd.
Printed and bound by
Clays Ltd, St Ives plc

Preface

In writing this book I have attempted to give a unified overview of the various aspects of the physics of soft condensed matter—including colloids, polymers, and liquid crystals—at a level suitable for an advanced undergraduate or a beginning graduate student. I hope that the book will be found of value by students of chemistry, materials science, and chemical engineering, as well as those students of physics doing one of the increasing number of courses about soft matter in today's physics curricula.

As a background, a student should have an elementary knowledge of the properties of matter. The theoretical underpinning of soft matter physics is in statistical mechanics and thermodynamics, and some acquaintance with the basics of these subjects is also desirable. An appendix is included summarising some key results. I am aware that many students find statistical mechanics rather difficult conceptually, and I have tried to choose transparency over elegance in deriving results. One additional advantage of studying soft matter is that it provides many illuminating examples of the principles of statistical mechanics at work. In particular the role of entropy is central; for this reason I have stressed methods for calculating thermodynamic quantities which rely on directly calculating the entropy, rather than the more compact but less transparent routes via the partition function.

Writing a textbook in an area that is not already well provided with them has both the advantage and the disadvantage that a traditional canon of topics does not yet exist. The choice of topics I have made reflects, of course, my own personal interests, and no doubt others would disagree about what should and should not be included. I hope that my choice, at least, will give students an appreciation both of the breadth of application of this fascinating class of materials and of the unity of some of the physical principles that underlie their behaviour.

The origins of this book lie in a final year undergraduate physics course taught at Cambridge. I owe a great debt to my colleagues who taught the course before me: Professors Athene Donald, John Field, and Mick Brown. This course, entitled 'Materials', covered structural aspects of many materials, both 'hard' and 'soft', though as the research interests of the department evolved the 'soft' element of the course became increasingly emphasised. When I moved to Sheffield I took the opportunity to use some of this material in writing a course for fourth-year MPhys students exclusively devoted to soft condensed matter, and it is this course that formed the basis for this book.

I am grateful to my colleagues Mark Geoghegan and Martin Grell, at Sheffield, and Joe Keddie, at Surrey University, who read the manuscript and pointed out a number of errors and potential improvements. Jon Howse kindly supplied the cover image. I would also like to thank Dr Sönke Adlung and his colleagues at Oxford University Press for the combination of enthusiasm and professionalism which they brought to the project.

Most of all, I thank my parents, Robbie and Sheila, and my wife Dulcie. They gave me very practical assistance during the preparation of the manuscript, as well as supporting and encouraging me throughout the period of its writing.

Sheffield   R.A.L.J.

Contents

Introduction and overview

1.1 What is soft condensed matter?

Soft condensed matter (or soft matter, for brevity) is a convenient term for materials in states of matter that are neither simple liquids nor crystalline solids of the type studied in other branches of solid state physics. Many such materials are familiar from everyday life—glues, paints, and soaps, for example—while others are important in industrial processes, such as the polymer melts that are moulded and extruded to form plastics. Much of the food we eat can be classed as soft matter, and indeed the stuff of life itself shares the qualities of mutability and responsiveness to its surroundings that are characteristic of soft matter. We are ourselves soft machines, in William Burroughs' apt phrase, and the material we are made of is soft matter.

In more precise terms, the materials we are discussing include colloidal dispersions, where submicrometre particles of solid or liquid are dispersed in another liquid, polymer melts or solutions in which the size and connectivity of the molecules lead to striking new properties, such as viscoelasticity, which are very different to those of a simple liquid, and liquid crystals, where an anisotropic molecular shape leads to states with a degree of ordering intermediate between a crystalline solid and a liquid.

What do these apparently disparate materials have in common? There are a number of features that they share which makes it worth considering them as a class. These include:

- The importance of length scales intermediate between atomic sizes and macroscopic scales. Colloidal particles are typically less than a micrometre in size, polymer chains have overall dimensions in the tens of nanometres, and the self-assembled structures formed by amphiphilic molecules have dimensions in a similar range. From the point of view of constructing theories, this means that one can (and should) use coarse-grained models that do not have to account for every detail on the atomic scale. These coarse-grained models emphasise universality; for example, many aspects of the behaviour of polymers derive not from the particular chemical details of the units that make up the chain, but simply from the topological implications that follow from the fact that the polymer molecule is a long, flexible curve in space which cannot be crossed by other chains.

- The importance of fluctuations and Brownian motion. Although typical structures in soft matter are larger than atomic sizes, they are small enough for Brownian motion—the fluctuations that take place in any thermal system—to be important, and the typical energies associated with the bonds between structures and with the distortions of those structures are comparable in size to thermal energies. Soft matter systems should be visualised as being in a constant state of random motion; polymer chains in solution are continually writhing and turning, while the membranes formed by sheets of self-assembled amphiphilic molecules are not rigid plates, but are continually buckling and flexing under the influence of Brownian motion.

- The propensity of soft matter to **self-assemble**. Related to the importance of Brownian motion is the fact that most soft matter systems are able to move towards **equilibrium**. But the equilibrium state of lowest free energy in a soft matter system is often not a state of dull uniformity; the subtle balances of energy and entropy in soft matter systems yield rich phase behaviour in which complex structures arise spontaneously. This **self-assembly** can take place at the level of molecules, but even more complexity occurs when ordering takes place **hierarchically**, with molecules coming together to form supramolecular structures (such as micelles), which themselves order at a higher level. In this way structures of tremendous intricacy and complexity are put together without external intervention, driven solely by the second law of thermodynamics.

We will see these themes recurring throughout the book.

1.2   Soft matter—an overview

The basic aim of condensed matter physics is to understand the collective properties of large assemblies of atoms and molecules in terms of the interactions between their component parts. In this book we will mostly be concerned with the structural and mechanical properties of soft matter, and the tools we will need are those of **statistical mechanics**. In Chapters 2 and 3 we review some of the concepts we will need, covering some material that will be familiar to many readers from elementary courses on the properties of matter, and introductory courses on solid state physics and thermal physics. At the macroscopic level, we need to be able to characterise the typical mechanical responses of solids and liquids; these responses can be understood at the microscopic level in terms of the role of bond energies and timescales for atomic or molecular motion. Elementary science stresses the distinction between solids and liquids, but we find cases that seem to stretch this definition: **viscoelastic** liquids, which seem to behave either like liquids or solids depending on the timescale at which they are probed, and **glasses**, which combine a liquid-like lack of long-ranged order with solid-like mechanical properties. **Phase behaviour** and **changes of phase** recur throughout the book; in Chapter 3 methods for treating both the equilibrium phase behaviour and the kinetics of phase transitions are introduced in the context of the unmixing of simple liquid mixtures. The methods introduced—in particular **mean field models** allowing one to calculate the free energy—are used again frequently in the book to deal with more complex systems. These two chapters thus provide a general framework in which the physics of more complex and specific systems covered in the later chapters can be dealt with in a unified way.

In Chapter 4, one specific class of soft matter is introduced—colloidal dispersions. This allows us to make some general points about the hydrodynamics of microscopic objects and Brownian motion, and gives one an opportunity to go into more detail about the forces that operate between surfaces at colloidal length scales. The discussion is mostly confined to the properties of hard, spherical, particles; the richness of behaviour found, for example, in suspensions of highly anisotropic particles such as clay platelets is not considered. However, even for spherical particles we find a wide range of behaviour: structural aspects include their assembly into colloidal crystals and aggregation into fractal structures, while in flow they display striking non-Newtonian effects such as shear thinning.

The properties of polymers are introduced in Chapter 5. Here we see a graphic illustration of the power of entropy in the random walk configurations taken up by long-chain molecules; understanding the role of this entropy allows us to understand the origin of the elastic properties of rubber. Combining this insight with an appreciation of how the topology of polymer molecules constrains their freedom to move allows us to construct a theory—**reptation**—that quantitatively accounts for the striking viscoelastic properties of polymer solutions and melts.

The formation of rubber from a melt of linear polymers involves **cross-linking**, and understanding this process, by which a liquid is turned into a material with a finite (but low) modulus, allows us to introduce in Chapter 6 another simple mathematical model that demonstrates how disparate physical systems can show a surprising degree of universality—**percolation**.

The next three chapters all consider aspects of **self-assembly** in soft matter. In Chapters 7 and 8, we consider self-assembly at the molecular level. In soft matter systems, we often find states of molecular order that are intermediate between the full three-dimensional order of a perfect crystal and the complete translational symmetry of a liquid. This intermediate order can be of two types. In Chapter 7 we consider **equilibrium** phases in which there may be only orientational order, or positional order only in one or two dimensions. These phases are known as **liquid crystalline** phases; some such materials are familiar to all of us as the basis of everyday display technologies for computers and calculators, but the variety of materials forming liquid crystalline phases is much wider, including a number of polymers. In Chapter 8 we consider crystallinity in polymers; here the reason for partial ordering is **kinetic** rather than thermodynamic. A fully crystalline state would have the lowest free energy, but such a state is inaccessible on experimental timescales, and it is the kinetics of the process of crystallisation which controls the intricate hierarchical structures that arise.

The theme of self-assembly is taken to another level in Chapter 9. Here we consider situations in which the units that come together to form ordered structures are not single molecules, but aggregates of molecules. The most familiar examples of such supramolecular self-assembly are found in soap molecules and similar amphiphiles, while analogous phenomena are to be found among polymers in **block copolymers**. Here liquid crystalline phases or even phases with full three-dimensional order are formed from units that are substantially bigger than molecular sizes.

The final chapter is rather different in its object to the others. Here I wish to introduce some of the areas of biology in which concepts drawn from the physics of soft condensed matter might prove of relevance. This chapter is much more tentative, but it seeks to highlight some important problems in molecular biology, such as the problem of **protein folding**, to which approaches from soft matter physics seem to offer promise. This chapter is inevitably incomplete and impressionistic, but with it I hope to be able to conclude the book with a sense of the potential of physics to make contributions in new and vital areas.

The constraints of brevity mean that many areas of soft condensed matter are treated only in rather a cursory way, while other areas (such as **foams**) are not covered at all. However, if I stimulate any reader to seek to read and think more deeply about any aspects of this fascinating branch of physics I will have achieved my aim.

Further reading

There are at present few books which cover the whole area of soft condensed matter physics. For a general introduction at a more descriptive level than this book, see Hamley (2000). Daoud and Williams (1999) is a collection of essays which give an excellent introduction to a number of aspects of soft condensed matter.

Forces, energies, and timescales in condensed matter

2

2.1 Introduction

Condensed matter is held together by **intermolecular forces**; the strength and range of these forces determine the bulk, macroscopic properties of matter. In a solid, each molecule is locked into a definite location on a crystal lattice as a result of these intermolecular bonds; there is a direct relationship between the **energy** of the bonds and the **stiffness** of the material. In a liquid, molecules are also held together by intermolecular forces in a state of high density, but in contrast to a solid the molecules are not locked rigidly into well-defined positions on a lattice. Instead, a molecule's position relative to its neighbours changes on a certain characteristic **relaxation time** which itself is determined by the way the random, thermal motion of the molecules is modified by the intermolecular forces. It is this relaxation time that determines how easily the liquid will flow when a stress is applied. In this chapter we will discuss the relationship between the bulk properties of matter, such as its response to an externally applied stress, and the forces acting between its component molecules.

One of our main tasks in this chapter is to establish the difference between solids and liquids. This difference is clear for many simple substances, but we shall see that in soft matter the distinction may be less clear-cut. Characteristic relaxation times for soft matter often fall into a range of values easily discernible to the human senses, and these materials seem to behave in a way that is neither entirely solid-like or entirely liquid-like—they are **viscoelastic**. In many systems, as the temperature is lowered the relaxation times increase dramatically, with the appearance of tending to infinity at some finite temperature. This divergence of the relaxation time leads to a non-equilibrium state of matter—a **glass**—in which a liquid-like lack of order is combined with solid-like elastic properties.

2.2 Gases, liquids, and solids

2.2.1 Intermolecular forces

If matter is made up of atoms and molecules, then the existence of condensed phases, such as solids and liquids, tells us something about the forces that act between the molecules. We know that there must be an attractive force which operates between molecules when they are separated; it is this force that has to overcome the effect of thermal agitation to cause a gas to condense to a liquid.

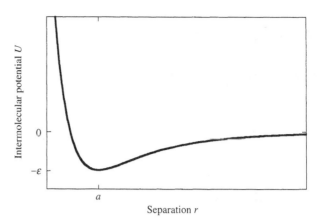

Fig. 2.1 Schematic plot of a typical inter-molecular potential, showing a short-ranged repulsion and a longer ranged attraction. The equilibrium separation is a, and the energy at the minimum is $-\epsilon$.

There must also be a repulsive force, which prevents matter from completely collapsing; the experimental observation that the compressibilities of liquids are rather high tells us that this repulsive force is rather short-ranged and strong. We can thus sketch the form of the potential; this is shown in Fig. 2.1. The force F may of course be derived from the potential by $F = -dU/dr$; at the minimum of the potential—the point of mechanical equilibrium—the force is zero. In this kind of plot we assume that the interaction between a pair of neighbouring molecules is isotropic, is unaffected by the interaction of either of the molecules with other molecules, and thus depends only on the separation of the pair of molecules in question. These assumptions are convenient, and indeed are almost essential for simple calculations, but as we shall see when we discuss the nature of the forces in more detail they are invalid for a number of important practical situations.

The origin of the short-ranged repulsion is essentially quantum mechanical; it arises as a result of Pauli's exclusion principle, which comes into operation when the electron orbitals of neighbouring molecules begin to interact. It is not straightforward to derive a simple closed expression for this potential, and the functional forms that are often used (usually exponentials or high inverse powers) are entirely empirical, and often chosen for calculational convenience rather than accuracy. In fact, for many purposes an extremely convenient model that captures much of the physics of liquids and solids assumes that the repulsive potential is essentially infinite as soon as the molecules overlap, but that there is a long-ranged attraction for larger separations. This idealisation is known as a **hard-sphere** potential with long-ranged interactions.

The nature of the long-range interaction depends on the system. Of course, all these interactions have a common origin in the electrostatic force, but it is convenient and traditional to distinguish between different types of interactions. These distinctions are probably familiar to readers from introductions to solid state physics and chemistry, but we recapitulate the key points here. Particularly important in the context of soft condensed matter is the order of magnitude of the bonds compared to thermal energies. If the bond energy is very much bigger than the thermal energy $k_B T$, then the probability of the bond breaking and reforming owing to thermal agitation is vanishingly small; these bonds can be thought of as **permanent** or **chemical** bonds. In contrast, if the bond energy is comparable to, or only a few times bigger than, $k_B T$, then there is a finite probability that the bond may be broken and subsequently reformed by thermal agitation; these bonds can be thought of as **temporary** or **physical** bonds.

- **Van der Waals forces.** These interactions arise between uncharged, weakly interacting atoms and molecules. All such atoms and molecules can be thought of as having a constantly fluctuating random dipole moment; this dipole will induce a corresponding dipole in a neighbouring atom or molecule leading to an attractive force. The potential U_{dis} between a pair of atoms or molecules, each with polarisability α, separated by a distance r, varies as

$$U_{\text{dis}} \sim \frac{\alpha^2}{r^6}. \tag{2.1}$$

(See Section 4.3.2 for a more detailed discussion.) The interaction is not strongly directional and typically the order of magnitude of the strength of the interaction between two molecules in contact is around 10^{-20} J. This is of the same order of magnitude as thermal energy $k_B T$ at room temperature.

- **Ionic interactions.** In an ionic solid transfer of electrons between atoms is essentially complete, and the resulting ions interact via a straightforward Coulomb potential U_C, which for ions carrying charges q_1 and q_2 at separation r is

$$U_C = \frac{q_1 q_2}{4\pi \epsilon_0 r}. \tag{2.2}$$

The interaction is non-directional, and is substantially stronger than the van der Waals interaction. A typical order of magnitude for an interaction between a pair of ions in a crystal is 100×10^{-20} J. This is a couple of orders of magnitude larger than thermal energy $k_B T$ at room temperature. The interaction is, however, strongly modified if the ions are in solution. The field due to a given ion is **screened**; because the other ions are free to move they can take up positions partially cancelling the field of the test ion. This is discussed in more detail in Section 4.3.3.

- **Covalent bonds.** In a molecule, electrons which originate from the component atoms interact with more than one nucleus, and the effect of this interaction is that the total energy of the molecule is lower than the energy of the separated atoms. This lowering of energy gives rise to an effective bonding between the atoms. Typical covalent bond energies are in the range of 30 to 100×10^{-20} J, very much greater than $k_B T$ at room temperatures. Covalent bonding is short-ranged and highly directional, and is not conveniently represented by simple interatomic potentials.

- **Metallic bonds.** A metallic bond is a particular variant of a covalent bond, in that it involves electrons being **delocalised** so as not to be associated with a single nucleus. However, whereas in a covalent bond the extent of delocalisation is limited to a pair, or at most a few nuclei, in a metal electrons are delocalised throughout a macroscopic volume of material. This delocalisation, as in the covalent bond, results in a reduction of the energy of the system. Metallic bonds are comparable in the magnitude of their bond energy to covalent bonds, but they are somewhat less directional.

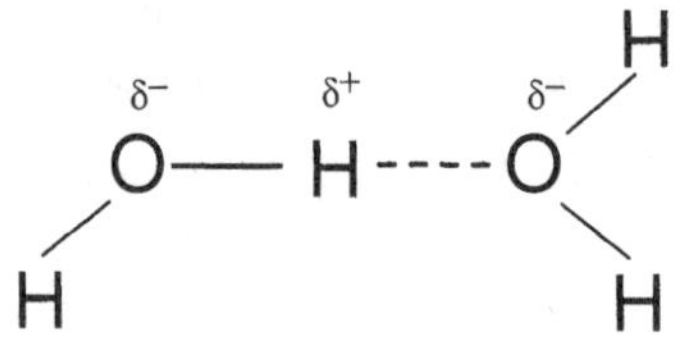

Fig. 2.2 A hydrogen bond between two water molecules.

- **Hydrogen bonds.** These are a special kind of interaction which occurs when a hydrogen atom is covalently bonded to an electronegative atom such as oxygen or nitrogen—see Fig. 2.2. Because of the hydrogen atom's small size and single electron, the side of the hydrogen atom which is furthest away from the electronegative atom presents a significant unshielded positive charge, which can interact with another electronegative atom. This results in an attractive energy intermediate in magnitude between a full covalent bond and a van der Waals interaction; typical values are between 2 and 6×10^{-20} J, corresponding to $25\text{--}100 k_B T$ at room temperature. Hydrogen bonding accounts for many of the peculiar properties of water, and also plays an important role in many biological macromolecules.
- **The hydrophobic interaction.** A consequence of the fact that liquid water forms a three-dimensional, hydrogen-bonded network is that molecules which cannot participate in hydrogen bonding with water perturb the local structure of liquid water around them. This perturbation of the local structure leads to a *decrease* in entropy—the water is locally made *more ordered* by the presence of the foreign molecule. This results in an increase in the free energy. The extra ordering imposed by two foreign molecules is reduced if the two molecules are brought closer together, leading to an effective attractive interaction. This attractive interaction is known as the **hydrophobic interaction**; the magnitude of the interaction is of order 10^{-20} J. This interaction is of crucial importance in promoting self-assembly in both biological and non-biological systems, and is discussed more fully in Chapter 9.

2.2.2 Condensation and freezing

The most obvious manifestation of the forces between molecules is the fact that on cooling a collection of molecules its physical state changes from a **gas** or **vapour** to a **liquid**, and on further cooling there is another change of state from a liquid to a **solid**. At high temperatures, the attractive forces between molecules are weak compared to the thermal energy; this thermal energy is present almost entirely in the form of kinetic energy. The molecules in the gaseous state are in a state of constant motion, interacting with each other only relatively infrequently through occasional collisions, and moving in a way which shows very little correlation between the motions of different molecules. This is the state that approximates to the familiar concept of the **ideal** or **perfect** gas. As the temperature is reduced, attractive interactions between the molecules start to become more important. On colliding, pairs of molecules stay together longer and correlations between the motion of different molecules start to appear, with short-lived clusters of molecules forming and breaking up. In this state, while the energy of the molecules is still dominated by the kinetic energy of their motion, the energies of interactions in the transient clusters start to be significant, and the properties of the gas start to deviate from the predictions of the perfect gas law. At some point the correlations between molecules become more permanent, and a new, dense phase of the material appears—the liquid. In the liquid state the attractive energy of interaction between the molecules is as important a part of the total energy as the kinetic energy of motion. The repulsive part of the interaction energy between the molecules also plays a role; the fact that two molecules cannot be in the same place at once leads to short-ranged correlations in the position of the molecules. The structure of the liquid is determined by the tension between the attractive part of the intermolecular potential, which tries to pack molecules as closely as possible, and the repulsive part of the potential, which imposes a minimum separation between the molecules. As the temperature is decreased further, another way of resolving this tension manifest itself. By packing the molecules together in a regular rather than random way, it is possible to achieve a higher density of molecules (and thus a larger contribution from the attractive part of the potential), while still satisfying the minimum distance constraint imposed by the repulsive part of the potential. The liquid has **frozen**.

The conditions under which solids, liquids, and gases form and coexist are summarised in a **phase diagram**. For a simple liquid, this plots the relationship between **temperature**, **pressure**, and **volume** (or equivalently, **density**). This relationship forms a surface in the three-dimensional space defined by these three variables; naturally it is conventional to represent this surface by a projection in two dimensions which can be easily plotted. The most familiar way of doing this is shown in Fig. 2.3(a), in which the regimes of temperature and pressure in which various phases are stable are plotted. In a typical experiment, one increases the temperature of the sample at a constant pressure. For low pressures, one has a direct transition between the solid and the vapour phase—the solid **sublimes**. At higher pressures, the solid first undergoes a transition into the liquid state—it **melts**—and then at a higher temperature the liquid is transformed into the vapour at a **boiling** point. At one particular pressure, there is a temperature at which all three phases, solid, liquid, and gas, coexist: this is the **triple point**. At another particular choice of pressure, the transition between the vapour and the liquid state becomes continuous rather than discontinuous; this combination of temperature and pressure is known as the **critical point**.

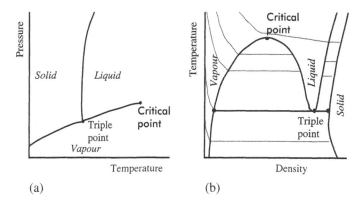

Fig. 2.3 Schematic phase diagrams for a simple elemental or molecular material. The familiar projection as a function of pressure and temperature is shown in (a), while (b) shows the same information plotted as density as a function of temperature. The dotted lines in (b) represent trajectories at constant pressure.

Figure 2.3(b) shows the same information plotted in a slightly less familiar way, with the phases given on a plane defined by temperature and density. This presentation emphasises the idea of **coexistence**; at a given pressure there is a temperature at which two out of the three phases will coexist. We will see parallels between this simple phase diagram for solid/liquid/gas coexistence and the more exotic phase diagrams that occur in soft matter later in the book.

2.3   Viscous, elastic, and viscoelastic behaviour

Normal condensed matter comes in two forms, solid and liquid. Typical 'soft condensed matter' seems more difficult to classify. Think of glue, soap, tomato ketchup, or pastes (e.g. cornflour and water)—are they solid or liquid? In some ways they seem to have attributes of both. In this section we will first review the ideal behaviour of normal solids and liquids, introducing the nomenclature and definitions we will need to use for more complex materials. Then we will discuss some of the more common types of non-ideal behaviour.

2.3.1   The response of matter to a shear stress

We can define a solid as a material that can sustain a shear stress without yielding while a liquid is something that flows in response to a shear stress. Before talking about the complexities of real materials, it is convenient to introduce idealisations of liquid and solid types of behaviour: the **Hookean** solid, characterised by perfectly elastic behaviour, and the **Newtonian** liquid, characterised by a single, time and shear rate independent, viscosity. Summarising these two types of behaviour, we can state that:

Hookean solid

- An applied shear stress produces a shear strain in response.
- The shear strain is proportional to shear stress, and the constant of proportionality is the shear modulus.

Newtonian liquid

- An applied shear stress produces a flow with a constant shear strain rate in response.
- The strain rate is proportional to the shear stress, and the constant of proportionality is the viscosity.

Before defining these idealisations mathematically, let us recall how shear stresses and shear strains are defined (see Fig. 2.4). The shear stress σ is given in terms of the applied force F and the area A as $\sigma = F/A$, while the shear strain e is given as $e = \Delta x/y$; alternatively, as one can see from the diagram, this is equivalent, for small strains, to the angle θ.

For a solid, the application of a stress to the material produces a constant strain in response; if the solid is a Hookean solid, the strain is simply proportional to the applied stress, with the constant of proportionality being the shear modulus G. Thus

$$G = \sigma/e. \tag{2.3}$$

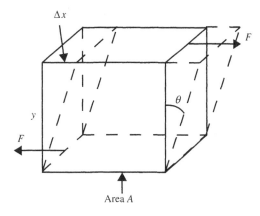

Fig. 2.4 Definitions for shearing a material.

For a liquid, a constant applied stress will result in a time-dependent strain. In a Newtonian liquid it is the strain rate which is constant when a constant stress is applied. In the elementary definition of viscosity, we imagine some liquid sandwiched between parallel plates of area A separated by a distance y. If the plates are moved with a relative velocity v, then the force F resisting the relative motion of the plates is given by

$$F = A\eta \frac{v}{y},\qquad(2.4)$$

where the coefficient η is the viscosity of the fluid. The velocity gradient v/y is in fact identical to the time derivative of the shear strain $\dot{e}$, so we can write this definition in a more general form as

$$\sigma = \eta \dot{e}.\qquad(2.5)$$

Thus we have defined two simple limiting behaviours for condensed matter: the elastic, or Hookean, solid, and the viscous Newtonian liquid.

Real materials—particularly soft matter—often behave in a way that combines viscous and elastic response, with an additional dependence on timescale. An example of this kind of behaviour is well seen in the material 'silly putty', available in toy shops. If one applies a stress on a slow timescale this material flows like a very viscous liquid, but if one rolls it up into a ball and drops it on a hard surface it bounces elastically. This type of behaviour is known as **viscoelasticity**; a viscoelastic material responds to an applied stress in a time-dependent way which is pictured in Fig. 2.5. Here we imagine a stress being applied at time $t = 0$ and held constant. The material responds at first in an elastic way, with a constant strain, but after a certain time τ it begins to flow like a liquid, with the strain increasing linearly with time. The time τ is the **relaxation time**; it is the time that separates the solid-like behaviour from the liquid-like behaviour. If a stress is applied on a timescale that is shorter than the relaxation time, the material will behave like a solid, as the silly putty does when we bounce it on the floor. For stresses that are applied on longer timescales than the relaxation time the viscoelastic material flows.

Inspection of Fig. 2.5 allows us to identify an important relation. If we define an instantaneous modulus G_0, which characterises the elastic response of the material at times much shorter than the relaxation time, and we characterise the

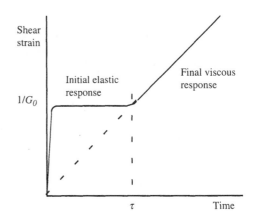

Fig. 2.5 Schematic strain response of a visco-elastic material to a shear stress applied at time $t = 0$ and subsequently held constant.

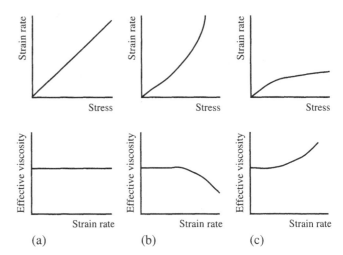

Fig. 2.6 Some possible responses of a fluid to an applied stress: (a) Newtonian; (b) shear thinning; and (c) shear thickening.

viscous behaviour at long times by a viscosity η, we can write an approximate relationship between G_0, η, and the relaxation time τ:

$$\eta \sim G_0\tau. \tag{2.6}$$

This relation may not be directly applicable to many real materials, because their viscoelastic response may be much more complicated than that shown in Fig. 2.5, with a time dependence that may be characterised by more than one single relaxation time. However, it is always at least dimensionally correct and it offers a useful conceptual framework for relating the phenomenology of viscoelastic materials to their response at a molecular level.

Another way in which complex fluids can display departures from ideal, Newtonian fluid behaviour is in having an effective viscosity which depends on shear rate. Then, for a steady flow, we can generalise eqn 2.5 to

$$\sigma = \eta(\dot{e})\dot{e}, \tag{2.7}$$

where $\eta(\dot{e})$ is a shear-rate dependent viscosity.

Possible responses of a fluid to an applied stress are sketched in Fig. 2.6. In contrast to a Newtonian fluid, a **shear-thinning** fluid becomes progressively easier to make flow as the shear rate gets larger. This is, for example, a desirable behaviour for a paint, which is easy to apply at the relatively high shear rates achieved by brushing, but as a thin film applied to a vertical surface does not significantly sag under its own weight. A **shear-thickening** fluid, on the other hand, flows relatively easily when a low shear rate is applied, but becomes much more resistant to flow when sheared at a high rate. This behaviour is often seen in pastes with rather a high volume fraction of particles. In fact, as we shall see in Chapter 4, both types of non-Newtonian behaviour are characteristic of concentrated dispersions of particles, and can be understood as being a consequence of the rearrangement of the particles in response to the flow.

2.3.2 Understanding the mechanical response of matter at a molecular level

When we impose a deformation on a solid we must impose a force; this is a direct consequence of the forces that hold the atoms of the solid in their equilibrium positions. Thus we can relate the elastic properties of the solid— and in particular the values of the various elastic moduli—to the intermolecular force curves between the component atoms in the solid.

This kind of calculation is simple in principle, though some of the details may become involved. We illustrate it by a calculation of the **Young modulus**, which relates the tensile strain to an applied tensile stress. Consider the structure sketched in Fig. 2.7, in which atoms are placed on a simple cubic lattice with interatomic separation a. The interactions between the atoms are represented by springs. If we stretch the sample in one direction so the interatomic spacing increases from a to r, then the force on each spring may be written $F = k(r-a)$, where k is the effective spring constant for each interatomic bond. The area per spring is a^2, so the tensile stress is $k(r-a)/a^2$. The tensile strain on the sample is $(r-a)/a$, so the Young modulus E is given by $E =$ (tensile stress)/(tensile strain) $= k/a$.

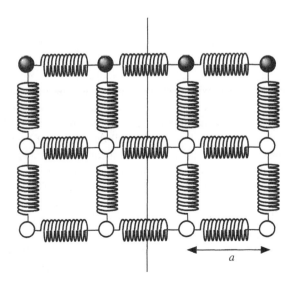

a

Fig. 2.7 Schematic model of a simple cubic solid for the calculation of the Young modulus of a solid in terms of intermolecular forces.

To determine the spring constant k in terms of the interatomic potential $U(r)$, we note that $U(r)$ can be expanded around the equilibrium separation a as

$$U(r) = U(a) + \frac{1}{2}(r-a)^2 \left.\frac{d^2U}{dr^2}\right|_{r=a} + \text{higher order terms}$$

$$= \text{constant} + \frac{1}{2}k(r-a)^2, \tag{2.8}$$

so we find

$$k = \left.\frac{d^2U}{dr^2}\right|_{r=a}. \tag{2.9}$$

To be general, and avoid having to make assumptions about the form of the interatomic potential, let us write it in the form

$$U(r) = \epsilon f(r/a), \tag{2.10}$$

where the potential has a minimum value $-\epsilon$ at $r = a$; that is to say, we can think of ϵ as a bond energy. The function $f(x)$ is dimensionless, with a minimum of -1 at $x = 1$. In terms of this function we find

$$k = \left.\frac{d^2U}{dr^2}\right|_{r=a} = \frac{\epsilon}{a^2}f''(1). \tag{2.11}$$

Here $f''(1)$ is simply a dimensionless number whose precise value depends on the precise functional form of the potential; let us simply write $f''(1) = A$, in terms of which we find for the Young modulus E

$$E = A\frac{\epsilon}{a^3}. \tag{2.12}$$

Thus, leaving aside the details, what we find from this kind of calculation is that an elastic modulus of a solid is related to the energy of a bond between neighbouring atoms or molecules multiplied by the density of those bonds. That a modulus is related to an energy density is clear from dimensional analysis; the modulus has dimensions of force divided by area, which is dimensionally equivalent to energy divided by volume. Thus a material with a high density of strong bonds is stiff, while a material with a low density of weak bonds is floppy.

The advantage of thinking about elastic moduli in terms of energy densities is that it makes it clear why even materials that flow in a viscous manner over long timescales can have finite values of the instantaneous modulus G_0. Imagine a liquid at some instant in time; its atoms will be arranged without any long-range order, but with a reasonably well-defined set of nearest neighbour distances. One can imagine applying a fixed shear strain to such an arrangement in such a way that the displacement of each atom relative to the other atoms reflects the macroscopic distortion material as a whole. This distortion raises the energy of the sample; the material is, temporarily at least, sustaining a finite shear stress, thus allowing us to define an instantaneous modulus.

The deformed material, now in a higher energy state than its equilibrium condition, seeks to relax to a lower energy, unstressed, state. This is where the difference between solids and liquids becomes apparent. In a perfectly ordered crystalline solid there is no mechanism available on any timescale for the stress to relax, because to produce a reduction in stress would require the wholesale rearrangement of atoms. In a liquid, on the other hand, motions of individual atoms are possible which reduce the stress without the need for wholesale, collective, motion of atoms.

The type of motion in a liquid which does reduce the stress is sketched in Fig. 2.8(b); here an atom (shaded darker than its neighbours) is in a high energy configuration, in which it is locally trapped by its neighbours, which together confine it to a 'cage' (shown by lines connecting the caging atoms). It is able to escape from its cage by a thermally activated jump into a new location (shaded light) of lower energy. The characteristic time for such a jump is the relaxation time τ; we can estimate it using an argument due originally to Eyring. The atom will be continuously vibrating within its cage, with some characteristic frequency ν. There is an energy barrier that must be overcome in order to escape from the cage; if this energy barrier is ϵ then the probability of the atom overcoming this barrier and escaping from its cage on any attempt is given by the Boltzmann distribution. Thus we can estimate the characteristic time to escape from the cage as

$$\tau^{-1} \sim \nu \exp\left(-\frac{\epsilon}{kT}\right). \tag{2.13}$$

How can we estimate the quantities that appear in this expression? The frequency ν must be comparable to the frequency of vibrations in a solid in which the interatomic forces are of similar magnitude.[1] To give us a feel for the order of magnitude of the barrier height ϵ, we note that an upper bound must be provided by the latent heat of vaporisation per molecule ϵ'. In fact, as we shall see when we look at the experimental results for the temperature dependence of viscosity, we find that for a wide range of liquids $\epsilon \approx 0.4\epsilon'$. Thus we find that relaxation times τ take values of order 10^{-12}–10^{-10} s in simple liquids. It is because these times are so short compared to typical experimental times that simple liquids apparently show purely viscous behaviour. In contrast, materials like polymer melts can have relaxation times of order milliseconds or even seconds, so their viscoelastic properties can be very spectacular. The mechanisms by which the tiny relaxation times characteristic of simple liquids are multiplied by many orders of magnitude in polymer melts to yield this impressive behaviour are discussed in Chapter 5.

Putting together these predictions, and recalling eqn 2.6, we find that we expect the viscosity η of a simple liquid to be given by an expression of the form

$$\eta = \frac{G_0}{\nu} \exp\left(\frac{\epsilon}{kT}\right). \tag{2.14}$$

The temperature dependence is dominated by the exponential, resulting in characteristic **Arrhenius** behaviour, the experimental signature of which is a straight line when the logarithm of viscosity is plotted against inverse temperature. Experimental results for many liquids confirm this picture, though as we shall see in the next section substantial deviations are found at low temperatures. This much stronger dependence of relaxation times on temperature is a result of the breakdown of the simple single-atom picture of relaxation, and leads to an entirely new phenomenon—the transition from a liquid to a glassy state.

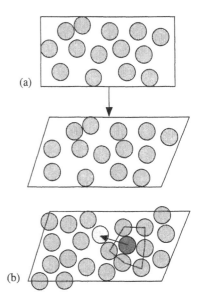

(a)

(b)

Fig. 2.8 (a) Instantaneous response of a liquid to an applied stress; the positions of the atoms are distorted without their relative arrangement changing. (b) Relaxation of stress in a strained liquid by the thermally excited escape of an atom (shaded dark) from the 'cage' imposed by neighbouring atoms to another site (shaded light).

[1] To be more specific, ν must be comparable to the frequency of phonons near the Brillouin zone boundary of a solid in which interatomic forces are of similar magnitude.

Before moving on to discuss glasses, it is worth noting in passing that while a solid with perfect crystalline order cannot flow on any timescale because its relaxation times are infinitely long, real solids can flow if a constant stress is applied to them. This mode of deformation is known as **creep**, and is of considerable importance in a number of engineering applications, particularly when solids are used at elevated temperatures. Creep is a result of the fact that real solids are not perfect crystalline lattices, but contain various defects, such as dislocations, grain boundaries, and vacancies. It is relaxation mechanisms involving these defects that are responsible for creep, and in very demanding applications where creep must be minimised, for example in turbine blades for jet engines, this can be done by using materials of maximum possible crystalline perfection—single crystals with the lowest possible density of dislocations.

2.4   Liquids and glasses

We saw in the last section that in simple liquids relaxation times are tiny compared to laboratory timescales. The temperature dependence of the relaxation time, and thus of the viscosity, is exponential, with an activation energy related to the latent heat of vaporisation per atom or molecule. This temperature dependence is significant, in that changing the temperature a few tens of degrees makes a significant difference to the viscosity, but it is not so strong that one would expect that lowering the temperature would cause the relaxation time to increase towards values comparable to experimental times. Surprisingly, for reasons that are not yet totally clear, this is exactly what happens for many, and possibly in principle all, liquids, both simple and complex, if they can be cooled to a low enough temperature without first crystallising.

2.4.1   Practical glass-forming systems

The range of materials that can form glasses is very wide, and as a result we all encounter a wide variety of glasses very frequently in everyday life. Some of the most important categories are listed below.

- Oxide glasses. Many oxides of elements from groups III, IV, and V of the periodic table, as well as mixtures of these and metallic oxides, can form glasses. The most familiar example of this class is ordinary window glass, a mixture with a typical composition 70% SiO_2, 20% Na_2O, and 10% CaO. These glasses are extensively used for structural and optical purposes.
- Chalcogenide glasses. The higher elements in group VI—sulphur, selenium, and tellurium, are known as **chalcogens**, and compounds of these elements with elements such as arsenic and germanium readily form glasses known as **chalcogenide glasses**.
- Elements. Three elements, sulphur, phosphorus, and selenium, readily form glasses in their pure form.
- Organic molecules. Many organic molecules, including solvents such as toluene and methanol, and medium molecular weight compounds such as glycerol and sucrose, as well as solutions of such molecules in water or organic solvents, form glasses. Glasses formed from sucrose solutions are familiar as boiled sweets or hard candy.

- Polymer glasses. Almost all polymers form glasses; polymers which are glassy in everyday used include poly(methyl methacrylate) and polycarbonates, which are extensively used for their transparency and good mechanical properties.
- Metallic glasses. Some metal alloys can form a glass if they are cooled very quickly, for example by directing a jet of liquid metal at a rapidly spinning, cooled copper disk. Cooling rates greater than 10^6 K s^{-1} can be obtained in such melt-spinning processes; the resulting metallic glasses have unique magnetic and mechanical properties that find applications in recording heads and transformer cores.

2.4.2 Relaxation time and viscosity in glass-forming liquids

While at relatively high temperatures it is found that relaxation times in liquids, and consequently their viscosities, do depend on temperature according to the Arrhenius law, at lower temperatures, assuming that the liquid has not first crystallised, the situation is different. The relaxation time associated with changes in configuration of the atoms or molecules, τ_{config}, assumes a temperature dependence that strongly departs from the temperature dependence of the characteristic time of vibrations τ_{vib}. This is sketched in Fig. 2.9; what is found is that the configurational relaxation time τ_{config} appears to diverge at a finite temperature T_0, the **Vogel-Fulcher temperature**. Experimentally, it is found that the temperature dependence of the relaxation time, and thus the viscosity, follows an empirical law known as the **Vogel-Fulcher law**:

$$\eta = \eta_0 \exp\left(\frac{B}{T - T_0}\right). \tag{2.15}$$

In practice, as the temperature is lowered, we reach a state at which the relaxation time becomes comparable to the timescale of the experiment τ_{exp}. When this happens, the system falls out of equilibrium with respect to configurational degrees of freedom. This marks the onset of the experimental glass transition, at a temperature T_g. When a liquid goes through a glass transition, it forms a **glass**.

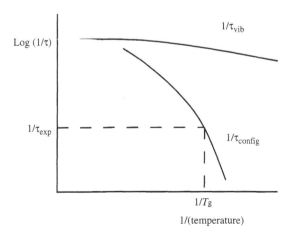

Fig. 2.9 Temperature dependence of relaxation times in a glass-forming liquid.

A glass is a material which is identical in its state of order with a liquid—that is to say, it has some short-range order but no-long range order, but behaves mechanically like a solid—and it has a finite shear modulus but an effectively infinite viscosity. It is quite misleading to state, as is often done by elementary textbooks, that a glass is simply a liquid with a very large viscosity. The appearance of a non-zero shear modulus is a qualitative change of properties, and indeed we find that the transition between a glass and a liquid is a sharp one, reminiscent in many ways of a phase transition, though it does have the fundamental difference that it is kinetic rather than thermodynamic in character.

2.4.3 The experimental glass transition

The transition between a liquid state and a glass is marked by discontinuities in thermodynamic quantities that are second derivatives of a free energy. If we follow the heat capacity or thermal expansivity of a glass-forming liquid we find that they change discontinuously at a temperature which is well defined for a given set of experimental conditions. This temperature is the **glass transition temperature** T_g.

For example, if we measure the volume of a glass-forming liquid as a function of temperature, the type of result we can expect is sketched schematically in Fig. 2.10. If the liquid is able to form a crystal, it will follow the path marked 'Crystal'; at some well-defined temperature T_m the volume will change discontinuously as the liquid forms a thermodynamically distinct crystal phase which has a qualitatively different degree of symmetry to the liquid. This is a first-order phase transition.

However, in some circumstances it is possible to cool a liquid down below its freezing point without it crystallising. This may be because the cooling rate is so fast that the liquid does not have time to crystallise, or it may be because the molecules have some permanent disorder which inhibits them from forming crystals at all. Whatever the reason, if the liquid does not crystallise it will follow a path such as that marked 'Glass (1)'. At some temperature below

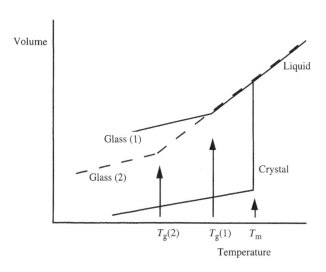

Fig. 2.10 Volume changes on cooling a glass-forming liquid.

the freezing temperature there is a change in slope of the graph of volume against temperature, corresponding to a discontinuous change in the thermal expansivity. This marks the glass transition temperature T_g. The glass transition is similar in appearance to a second-order phase transition, but it is not a true thermodynamic phase transition. This is because the transition temperature depends on the **rate** at which we do the experiment.

To see this, we need to revisit Fig. 2.9. When the timescale that characterises our experiment, τ_{exp}, is equal to the timescale characterising the rearrangement of the positions of molecules with respect to each other, τ_{config}, the degrees of freedom associated with the translational motion of the molecules can no longer come to equilibrium on the timescale of the experiment. Because of this, these degrees of freedom can no longer contribute to the values of thermodynamic quantities as they are measured in the experiment, and we observe a glass transition.

If now we do our experiment more slowly, we reduce τ_{exp}, and so we can go to a lower temperature before $\tau_{exp} \sim \tau_{config}$. Thus at lower cooling rates we will observe a curve like that marked 'Glass (2)' in Fig. 2.10, with a lower glass transition temperature $T_g(2)$.

Experimentally, it is now more usual to detect the glass transition using calorimetry. If we measure the heat capacity c_p (measured at constant pressure) of a glass-forming liquid, we find curves of the kind sketched in Fig. 2.11. The experimental glass transition is marked by a discontinuity in heat capacity; exactly as is found in measurements of volume, the value of experimental glass transition depends on the heating rate at which the experiment was carried out.

The heat capacity at constant pressure, c_p, is related to the **entropy** of the system by the relationship

$$c_p = T\left(\frac{\partial S}{\partial T}\right)_p. \tag{2.16}$$

Thus we can find the entropy as a function of temperature by integrating experimental heat capacity curves. The result is sketched in Fig. 2.12, together with the behaviour expected if the liquid freezes at a melting temperature T_m. Qualitatively, the features are very similar to those observed in a plot of volume against temperature; for liquids that are undercooled below their melting point,

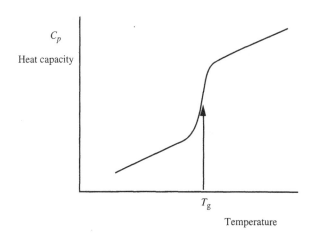

C_p

Heat capacity

T_g

Temperature

Fig. 2.11 Measured heat capacity as a function of temperature in a glass-forming liquid. The apparent discontinuity in heat capacity marks the experimental glass transition; its value depends on the heating rate at which the experiment is carried out.

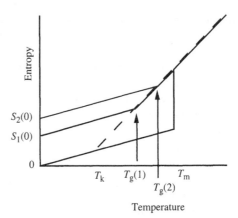

Fig. 2.12 Entropy changes on cooling a glass-forming liquid. $T_g(2)$ represents the experimental glass transition observed using a faster cooling rate than $T_g(1)$.

the entropy continues to decrease. At the experimental glass transition, the variation of entropy with temperature changes slope. As before, the location of the experimental glass transition depends on the rate at which the experiment is done; if we cool the liquid more slowly, we obtain a lower value for the glass transition temperature.

The variation of entropy with temperature in a glass-forming liquid presents a number of interesting and instructive features, which we summarise as follows.

- When extrapolated to absolute zero, the entropy of the glass remains finite. The glass has a certain **residual entropy** associated with its configurational disorder.
- The entropy of a glass defined in this way is not a pure thermodynamic function of state, because it depends not only on the pressure and temperature, but also on the **history** of the sample. This means that in the glassy state, the system is not able to sample all the possible statistical mechanical microstates on an experimental timescale. This situation is sometimes called **broken ergodicity**.
- If we imagine doing an experiment in which the cooling rate is successively decreased, we expect lower and lower measured glass transitions. But there must come a temperature at which the continuation of the entropy curve for the supercooled liquid intersects the entropy curve for the crystalline solid. Since it does not make physical sense for the glass to have a lower entropy than the crystal, it is suggested that this temperature, which is known as the **Kauzmann temperature** T_K, marks an effective lower limit on the experimental glass transition temperature. Estimates of the Kauzmann temperature reveal that it is often remarkably close to the Vogel–Fulcher temperature T_0.

Experimentally, then, we see that the glass transition has some features in common with a second-order phase transition: quantities like the expansivity and the heat capacity (which formally can be written in terms of second derivatives of the free energy) are discontinuous at the transition. However, despite this superficial similarity the glass transition is not a phase transition. The location of the transition depends on the rate at which the experiment is carried out, and the glassy state is not an equilibrium phase, because it is not the state of lowest free energy. Instead, we should understand the glass transition as a **kinetic** transition; it is the point at which the system is no longer able to sample all the microstates on an experimental timescale. Thus it is the divergence of the relaxation time in glass-forming liquids that is at the root of the glass transition phenomenon.

2.4.4 Understanding the glass transition

Glasses are ubiquitous and familiar materials, so it is rather sobering to appreciate that even now there is no wholly satisfactory theory to explain the formation and behaviour of glasses. Instead, there are a variety of different theoretical approaches, which, while they do not add up to a consistent and complete picture, do give us some insight into the origins of the glassy state.

A popular early approach to glass formation was provided by **free volume theory**, which is still described in many textbooks (particularly of polymer science and materials science) despite a number of serious shortcomings. These theories start from the assumption that liquids contain a certain amount of **free volume**, which is available to permit motion of nearby segments. If we write the total volume of the sample as v, and the free volume v_f, then it is argued that the fractional free volume v_f/v can be written as a simple linear function of temperature, as

$$\frac{v_f}{v} = f_g + \alpha_f(T - T_g), \tag{2.17}$$

where f_g is the fractional free volume at the glass transition temperature T_g, and α_f is the expansion coefficient of the free volume. If there is a simple relationship between free volume and the viscosity η of the form

$$\eta = a \exp\left(\frac{bv}{v_f}\right), \tag{2.18}$$

where a and b are constants, and v is the physical volume of the sample, then the Vogel–Fulcher relationship can be recovered.

In spite of this success, the theory has two major failings. Firstly, the rather arbitrary assumptions which underlie eqn 2.17 mean that the theory does not have a great deal of predictive power, and in any case these assumptions are open to serious question. Secondly, some new predictions that the theory does make are not borne out by experiment. In particular, by varying the temperature and pressure simultaneously it is possible to make a polymer system go through a glass transition in a state of constant free volume, which is difficult to account for in this theory.

A physical idea that has proved useful in constructing theories of the glass transition is the idea of **cooperativity**. This is illustrated in Fig. 2.13. At high temperatures and relatively low densities, the space that is required for a molecule to make a move out of its immediate location can be made available simply by the uncoordinated, local vibrational motion of its neighbours. At lower temperatures and densities, the local motion of its neighbours is not sufficient to allow a molecule to move. Instead, a number of its neighbours must move **cooperatively** in order to make space. This idea was made concrete by Adam and Gibbs (1965); they introduced the concept of a **cooperatively rearranging region** to describe the minimum number of molecules that have to

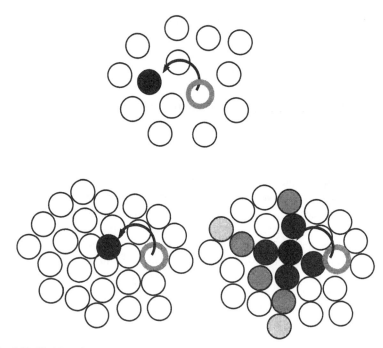

Fig. 2.13 The idea of cooperativity in diffusive motion of a molecule in a glass-forming liquid. At high temperatures and low densities (top) a molecule is able to jump to a new position without the necessity for wholesale rearrangement of its neighbours. At lower temperatures and higher densities (bottom), in order for one molecule to be able to make a move (left), some of its neighbours (shown shaded) must move cooperatively to make room (right).

move cooperatively in order for any motion to take place. As the temperature is lowered, the size of this region becomes larger, becoming infinite at the Vogel–Fulcher temperature, at which the viscosity diverges.

Supposing the energy barrier for a single molecule to move is $\Delta\mu$, and the number of molecules in the cooperatively rearranging region is z^*, then we assume that the activation barrier for motion in the cooperatively rearranging region is simply proportional to the number of molecules that have to move, z^*. Thus we can write for a relaxation time

$$\tau^{-1} \sim \nu \exp\left(-\frac{z^*\Delta\mu}{kT}\right), \tag{2.19}$$

with ν a microscopic frequency. Thus in this picture the non-Arrhenius behaviour of the relaxation time of a glass is simply a consequence of the larger number of molecules that have to move cooperatively as the temperature is decreased. Adam and Gibbs further argued that the size of the cooperatively rearranging region is inversely related to the **excess configurational entropy** of the liquid S_c—that is to say, that part of the entropy associated with the freedom of the molecules to adopt new configurations. If we refer to Fig. 2.12, we can picture this quantity as the difference between the entropy of the supercooled liquid and the entropy of the crystalline solid. At the Kauzmann temperature, this quantity goes to zero; we identify this temperature with the Vogel–Fulcher temperature at which the size of the cooperatively rearranging region diverges. This gives us an equation for the relaxation time

$$\tau^{-1} \sim \nu \exp\left(-\frac{C}{T S_{\mathrm{c}}}\right), \qquad (2.20)$$

where C is a constant. We can obtain an experimental determination of S_{c} from the difference in measured heat capacities between the supercooled liquid and crystal state ΔC_p; often this is found to be inversely proportional to temperature. If this is the case it can be shown from expression 2.20 that the viscosity should follow the Vogel–Fulcher form.

Further reading

For a good general introduction to the relationship between the macroscopic properties of matter and the forces between their constituent atoms and molecules, see for example Tabor (1991).

More details on glasses can be found in Zarzycki (1991), which concentrates on inorganic glasses, and Elliott (1990), which also deals extensively with the electronic and optical properties of inorganic materials. An in-depth review of theories of the glass transition can be found in Jäckle (1996).

Exercises

(2.1) A simple viscometer consists of two concentric cylinders, each of length $100\,\mathrm{mm}$. The inner cylinder, with radius $95\,\mathrm{mm}$, rotates at an angular velocity of $5\,\mathrm{rad\,s}^{-1}$, while the outer one, with radius $100\,\mathrm{mm}$, is fixed. The viscometer is filled with an oil whose viscosity is 10 poise. What torque is measured on the outer cylinder?

(2.2) A similar apparatus to the one in the last question has the gap between the cylinders filled with a rubber with a shear modulus of $5 \times 10^5\,\mathrm{Pa}$. The inner cylinder is rotated from the rest position through an angle of 0.01 radians. What is the torque on the outer cylinder?

(2.3) Consider a solid with a simple cubic structure, for which the interatomic potential $U(r)$ is written

$$U(r) = \frac{A}{r^n} - \frac{B}{r^m}.$$

Having first derived expressions for the equilibrium separation a and the energy at equilibrium $-\epsilon$, derive an expression for the Young modulus in terms of the bond energy ϵ and the equilibrium separation a. Assume that when a tensile stress is applied, the interatomic separations in directions perpendicular to the stress remain unchanged.

(2.4) The data below give bond energies, densities, relative atomic/molecular masses, and Young moduli for a number of solids.

Element	Bond energy (kJ mol^{-1})	Density (g cm^{-3})	Rel. atomic mass (g mol^{-1})	Young modulus (N m^{-2})
Ar	7.74	1.77	40	4.74×10^9
Li	665	0.533	6.94	3.0×10^9
Na	601	0.971	23	1.7×10^9
NaCl	752	2.17	58.45	5.4×10^{10}
Diamond	710	3.514	12	9.0×10^{11}
Si	446	2.33	28	1.3×10^{11}

a) Calculate the bond energy per molecule and express it in units of $k_{\mathrm{B}}T$ where T is room temperature, $300\,\mathrm{K}$.

b) In each case, identify the type of bonding involved.

c) Use the data to test the suggestion that the Young modulus is proportional to the bond energy per molecule divided by the density of bonds.

(2.5) The table below gives the temperature dependence of the viscosity of water, whose latent heat of vaporisation is $226 \times 10^4 \, \mathrm{J\,kg^{-1}}$.

Temperature/ °C	0	10	20	30	40
Viscosity/ 10^{-4} Pa s	17.93	13.07	10.02	7.98	6.53

50	60	70	80	90	100
5.47	4.67	4.04	3.54	3.15	2.82

a) Use this data to test the relation given in Section 2.3.2 for the viscosity of a liquid.

b) Assuming that the instantaneous shear modulus of water, G_0, is comparable to the shear modulus of ice at $0\,°\mathrm{C}$, 2.5×10^9 Pa, estimate the characteristic frequency of vibration for water.

(2.6) For polystyrene, the variation of viscosity with temperature follows the Vogel–Fulcher law with parameters $B = 710$ and $T_0 = 50\,°\mathrm{C}$. Plot the function η/η_0 in the temperature range 80–150 °C. By what factor does the viscosity and relaxation time vary between the temperatures of 100 °C and 140 °C?

(2.7) For polystyrene, a relaxation time associated with configurational rearrangements, τ_{config}, follows a Vogel–Fulcher law,

$$\tau_{\mathrm{config}} = \tau_0 \exp\left(\frac{B}{T - T_0}\right),$$

where τ_0, B, and T_0 are constants, with $B = 710$ and $T_0 = 50\,°\mathrm{C}$. A value of the experimental glass transition temperature is measured with an experiment carried out at an effective timescale $\tau_{\mathrm{exp}} = 1000\,\mathrm{s}$ and found to be $101.4\,°\mathrm{C}$.

a) Another experiment is carried out at an effective timescale of 10^5 s. What is the value of the glass transition temperature that this experiment obtains?

b) Or what timescale must an experiment be carried out if it is to measure a glass transition temperature within 10 °C of the Vogel–Fulcher temperature T_0? Is this practically possible?

(2.8) From 0 to 152 K the heat capacity, C_p^{g}, of a particular glass as normally measured is essentially the same as that of the stable crystalline phase, C_p^{c}. Around 160 K, there is an experimental glass transition, and the heat capacity of the glass rises above the heat capacity of the crystal; at 162 K, $\Delta C_p = C_p^{\mathrm{g}} - C_p^{\mathrm{c}} = 180 \, \mathrm{J\,K^{-1}\,mol^{-1}}$, for temperatures from 162 K to the melting point of the crystalline phase $\Delta C_p \propto T^{-1}$, where T is the temperature.

a) Sketch the behaviour of C_p, the enthalpy H, and the entropy S.

b) How might the thermodynamic properties vary with the rate at which the glass was quenched from the melt?

c) If the heat of fusion of the crystalline form is $24.2\,\mathrm{kJ\,mol^{-1}}$, estimate a lower bound, T_0, for the glass temperature assuming that ΔC_p obeys a $1/T$ law between T_0 and the melting temperature, 237 K.

Phase transitions

3.1 Phase transitions in soft matter

Soft matter often has a very rich and complicated morphology; rather than having a simple structure its components can be arranged in a complicated way, involving features at length scales intermediate between the atomic and the macroscopic. Examples will be discussed in detail in later chapters, and include block copolymer morphologies, soap morphologies, and emulsions. The remarkable feature of these structures is that they put themselves together in very complicated arrangements without outside assistance—they **self-assemble**. Understanding self-assembly involves understanding phase transitions and their kinetics. This is important for soft condensed matter, but much of what we need to know applies also to simpler kinds of matter too, and our discussion and examples will not be restricted to soft matter.

There are two general classes of self-assembled structures: structures that are essentially at **equilibrium**, and **non-equilibrium** structures that occur following a phase change. In normal solids and liquids the equilibrium situation is usually rather uninteresting—a perfect crystal, or a mixture of two liquids macroscopically separated into two layers, for example—but soft matter systems can often exhibit complex and interesting equilibrium phases. In Chapter 9, we will discuss two good examples of this: in block copolymers and surfactant solutions the equilibrium morphology consists of structures on length scales intermediate between the molecular and the macroscopic.

Non-equilibrium self-assembled structures often occur following a **phase transition**. If we change some external parameter (e.g. temperature) the structure with the lowest free energy may discontinuously change its character. Thus a **qualitative** change of structure occurs in response to a **quantitative** change in a control parameter.

Phase transitions almost always involve a change from a more ordered state to a less ordered state. For example, a liquid is less ordered than a solid. Phase transitions reflect a change with temperature of the relative importance of entropy and energy. The balance between entropy and energy is reflected in the free energy; for example, for changes at constant volume the appropriate free energy is the Helmholtz free energy F, defined by $F = U - TS$, where U is internal energy and S is entropy. In general at high temperature disordered phases are more stable.

For a phase transition one can define an order parameter, which typically takes a zero value in the disordered phase, and a finite value in the ordered phase. The way the order parameter varies with temperature tells one about the nature of the transition. A fundamental distinction is between first-order phase transitions, where the order parameter changes discontinuously at the phase transition (e.g. melting of a crystal), and second-order transitions, where the order parameter is continuous. The classic example of a second-order transition is the change from a liquid to a gas at a critical point.

Information about which equilibrium phases have the lowest free energy is not sufficient by itself to explain all the types of structure one can obtain in soft matter, or indeed in other types of condensed matter. In addition, one needs to understand the kinetics of the process by which phase ordering or disordering proceeds. Just because at equilibrium a system will take up a certain minimum free energy structure does not mean that when one goes through a phase transition the system will instantly adopt its new structure. Atoms and molecules will have to rearrange themselves, and this motion takes time. This time can be very long, so that systems can be caught in states in which they have not yet reached equilibrium. In many cases interesting structures are formed in the process, and although these structures are not at equilibrium they may be effectively frozen in, for example by cooling liquids through a glass transition. We need to understand the mechanisms by which phase transitions take place, and the nature of the non-equilibrium structures that the system must go through on its way to reach equilibrium.

In this chapter we will illustrate these concepts by discussing two examples: the unmixing of two immiscible liquids, and melting and crystallisation. The concepts we introduce here will then be used to discuss more complex transitions of the kind important in soft matter.

3.2   Liquid–liquid unmixing—equilibrium phase diagrams

Our first example of a phase transition describes the situation when two liquids are miscible in all proportions at high temperature, but separate into two distinct phases when the temperature is lowered. We use this example to introduce a simple statistical mechanical model, which although oversimplified, captures most of the important physics while remaining mathematically relatively easy to handle. Using this model we can calculate the equilibrium state of the system as a function of temperature and composition—the phase diagram—and also, in the next sections, we can use it as a basis for theories of the mechanisms by which phase separation takes place. In later chapters similar models, with only slight alterations, can be used to understand phase behaviour and self-assembly in soft matter systems.

The model we use is the **regular solution model**. This is a **mean field theory** with strong family resemblances to other mean field theories in statistical physics, such as the Bragg–Williams theory for order–disorder transitions in alloys and the Curie–Weiss theory for the paramagnetic–ferromagnetic phase transition. Our aim is to be able to predict the **free energy of mixing**. This quantity is illustrated in Fig. 3.1; on the left the two species A and B exist in separate, unmixed states; we write the free energy as $F_A + F_B$. On the right the two species form a single, mixed phase, whose free energy we write as F_{A+B}. The free energy of mixing $F_{mix} = F_{A+B} - (F_A + F_B)$. If we can predict this

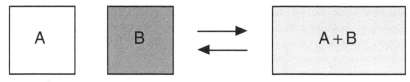

Fig. 3.1 The free energy of mixing. On the left, we have two species A and B existing in an unmixed state, while on the right, they form an intimate, molecular mixture. The free energy of mixing is the change in free energy going from one state to the other.

quantity as a function of composition and temperature we have everything we need to know about the phase behaviour of A and B.

In calculating F_{mix}, we need to calculate both the change of entropy on mixing S_{mix} and the change of energy on mixing U_{mix}. To find the entropy change, we imagine that the molecules of the two liquids are arranged on a lattice, where each lattice site has z nearest neighbours. (Of course, in liquids the lattice is not a regular crystalline one.) We will measure the composition of the mixture in terms of the **volume fraction** ϕ; the volume fraction of A molecules ϕ_A is defined as the volume of A molecules divided by the total volume of the system. For simplicity, we will assume that the total volume of the system is constant, independent of composition, so $\phi_A + \phi_B = 1$, and that the volume of an A molecule is the same as the volume of a B molecule.

If the volume fraction of A molecules is ϕ_A, and the volume fraction of B molecules is ϕ_B, we do not know for certain whether a given site is occupied by A molecules or B molecules. We can now use the Boltzmann formula to write down the entropy per site associated with this uncertainty, S_{mix}. The Boltzmann formula is

$$S = -k_B \sum_i p_i \ln p_i, \qquad (3.1)$$

where the sum is taken over all the states of the system i, each of which has the probability p_i. Here there are just two states: either the site is occupied by an A molecule or it is occupied by a B molecule, with probabilities ϕ_A and ϕ_B respectively. Thus

$$S_{mix} = -k_B (\phi_A \ln \phi_A + \phi_B \ln \phi_B). \qquad (3.2)$$

Here we have assumed that neighbouring sites are independent of each other, so that if one site is occupied by a B molecule it makes it neither more nor less likely that a neighbouring site is similarly occupied. This is a **mean field** assumption. We note that the entropy of mixing is zero if either ϕ_A or ϕ_B is unity, as we expect for a pure liquid.

This argument gives us the entropy of mixing—now we need to find the **energy** of mixing U_{mix}. To do this we assume that molecules interact only with their nearest neighbours in a way that is **pairwise additive**. We assume that the energy of interaction between two neighbouring A molecules is ϵ_{AA}, between two neighbouring B molecules ϵ_{BB}, and between an A molecule and a neighbouring B molecule ϵ_{AB}. Once again, we use a mean field assumption; we assume that a given site has $z\phi_A$ A neighbours, and $z\phi_B$ B neighbours, whether the site is itself occupied by A or B molecules. With this assumption, we can write the interaction energy per site as $(z/2)(\phi_A^2\epsilon_{AA} + \phi_B^2\epsilon_{BB} + 2\phi_A\phi_B\epsilon_{AB})$. If we subtract from this the energy of the unmixed state, $(z/2)\times(\phi_A\epsilon_{AA} + \phi_B\epsilon_{BB})$ we find for the energy of mixing

$$U_{mix} = \frac{z}{2}[(\phi_A^2 - \phi_A)\epsilon_{AA} + (\phi_B^2 - \phi_B)\epsilon_{BB} + 2\phi_A\phi_B\epsilon_{AB}]. \quad (3.3)$$

If we can assume that $\phi_A + \phi_B = 1$ (as will be the case if the mixture is incompressible), we can introduce a single dimensionless parameter, χ, which characterises the strength of the energetic interaction between A and B relative to their self-interactions. We define

$$\chi = \frac{z}{2k_B T}(2\epsilon_{AB} - \epsilon_{AA} - \epsilon_{BB}); \quad (3.4)$$

thus χ is the energy change in units of $k_B T$ when a molecule of A is taken from an environment of pure A and put into an environment of pure B. Using this definition we can write the energy of mixing in the simple form

$$U_{mix} = \chi\phi_A\phi_B. \quad (3.5)$$

Now, using the equation for free energy $F = U - TS$ we can write down the free energy of mixing as

$$\frac{F_{mix}}{k_B T} = \phi_A \ln\phi_A + \phi_B \ln\phi_B + \chi\phi_A\phi_B. \quad (3.6)$$

Now we have an expression for the free energy of mixing as a function of the concentration of the mixture that depends on a single dimensionless parameter χ, which expresses the strength of the energetic interaction between the components.

We can understand the phase behaviour of our mixture by looking at the way the shape of the curves of free energy against composition change with varying values of χ. Examples of these curves are shown in Fig. 3.2. For

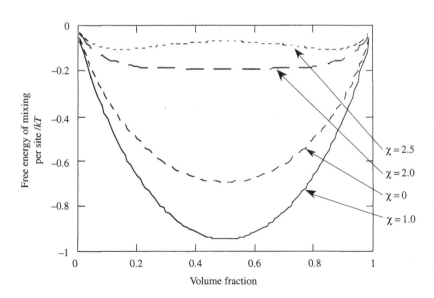

Fig. 3.2 The free energy of mixing divided by $k_B T$, as a function of composition for various values of the interaction parameter χ, as given by eqn 3.6 together with the incompressibility assumption $\phi_A + \phi_B = 1$.

negative values of χ, the curve has a single minimum at $\phi_A = \phi_B = 0.5$, while for values of $\chi \geq 2$ we find two minima, and a maximum at $\phi_A = \phi_B = 0.5$. To understand the physical significance of this, we must look at the two possible situations in more detail.

Suppose we have a volume V_0 of mixture whose starting volume fraction of A is ϕ_0. (From here on we will assume all volume fractions refer to A, and drop the suffix, i.e. we have $\phi_A = \phi$ and $\phi_B = 1 - \phi$). If this mixture separates into a volume V_1 with volume fraction ϕ_1 and a volume V_2 with volume fraction ϕ_2, then to conserve the total amount of A and B, $\phi_0 V_0 = V_1 \phi_1 + V_2 \phi_2$. Thus we can write

$$\phi_0 = \alpha_1 \phi_1 + \alpha_2 \phi_2, \tag{3.7}$$

where the relative proportions of the two phases α_1 and α_2 obey

$$\alpha_1 + \alpha_2 = 1. \tag{3.8}$$

The total energy of the phase-separated system is $F_{\text{sep}} = \alpha_1 F_{\text{mix}}(\phi_1) + \alpha_2 F_{\text{mix}}(\phi_2)$, which we can rewrite using eqns 3.7 and 3.8 as

$$F_{\text{sep}} = \frac{\phi_0 - \phi_2}{\phi_1 - \phi_2} F_{\text{mix}}(\phi_1) + \frac{\phi_1 - \phi_0}{\phi_1 - \phi_2} F_{\text{mix}}(\phi_2). \tag{3.9}$$

This equation is best interpreted graphically. In Fig. 3.3(a), an initial composition ϕ_0 phase separates into two phases ϕ_1 and ϕ_2. The total free energy of these two phases, F_{sep}, given by eqn 3.9, can be read off the straight line joining $F(\phi_1)$ and $F(\phi_2)$ where the volume fraction takes the value ϕ_0. For the concave curve shown in Fig. 3.3(a), the free energy resulting from phase separation into any pair of volume fractions ϕ_1 and ϕ_2 is always higher than the free energy of the starting composition F_0, so the mixture is stable.

On the other hand, if there is any region of composition in which the curve is convex, as shown in Fig. 3.3(b), then there are some starting compositions which, if they phase separate, can lead to a **lowering** of free energy. It is clear that the limiting compositions which bound this region of phase separation are those compositions joined by a common tangent, ϕ_1 and ϕ_2. These compositions are known as the **coexisting** compositions, and the locus of these compositions as the temperature, and thus the interaction parameter, is changed, is known as the **coexistence curve** or the **binodal**.

If we look in more detail at the free energy curve for compositions that fall within the the coexistence curve (see Fig. 3.4), we find another important distinction: the curvature of the free energy function $d^2 F/d\phi^2$ may be either positive or negative. At a composition ϕ_b phase separation to two compositions close to ϕ_b results in a lowering of free energy from F_b to F_b'. At this composition the system is unstable with respect to small fluctuations in composition, and will immediately begin to phase-separate. This composition is **unstable**. However, at composition ϕ_a a similar small change in composition leads to an increase in free energy from F_a to F_a'; the system is locally stable with respect to such small composition fluctuations, even though it is still globally unstable with respect to separation into the two coexisting phases. There is an energy barrier which needs to be surmounted in order to achieve the global energy minimum associated with phase separation, and as a result this

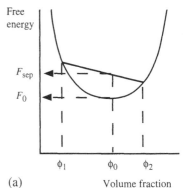

(a)   Volume fraction

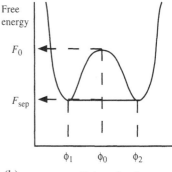

(b)   Volume fraction

Fig. 3.3 The free energy of mixing as a function of composition for one-phase and two-phase mixtures. (a) An initial composition ϕ_0 phase separates into two phases ϕ_1 and ϕ_2; the total free energy of these two phases, F_{sep}, is always higher than the free energy of the starting composition F_0 so the mixture is stable. (b) mixtures with compositions between ϕ_1 and ϕ_2 can lower their free energy by separating into two phases at these compositions.

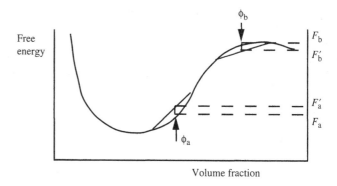

Fig. 3.4 The free energy of mixing as a function of composition, showing the distinction between compositions that are **metastable**, such as ϕ_a, and compositions that are **unstable**, such as ϕ_b.

composition is **metastable**. Clearly the limit of local stability is defined by the condition that the curvature $d^2F/d\phi^2 = 0$; the locus of these points is known as the **spinodal**.

Finally we note that a critical temperature T_c separates the two types of situation for which mixtures are always stable, and situations for which there are compositions which will phase separate. In the former, the curvature of the free energy function $d^2F/d\phi^2$ is always positive, while in the latter $d^2F/d\phi^2$ is negative within a certain range of ϕ. The critical point is thus defined by the condition $d^3F/d\phi^3 = 0$, and it is the point at which the coexistence curve and spinodal line meet.

Knowing these relationships between the shape of the free energy of mixing as a function of composition and the phase behaviour of the mixture allows us to calculate a **phase diagram** for the mixture—a plot that shows on a plane of composition and interaction parameter or temperature the regions where the mixture is stable, unstable, or metastable. These calculations are easily done for the regular solution model we have described above, with the free energy function given by eqn 3.6. A very useful simplification in this case follows from the fact that this function is symmetric about $\phi = 0.5$; in this case the condition for the double-tangent construction that defines the coexistence curve is simply $dF/d\phi = 0$.[1] The resulting phase diagram is shown in Fig. 3.5.

We see from this diagram that in any situation in which the interaction parameter χ is less than two, the mixture is completely miscible in all proportions. When the interaction parameter χ is less than zero, this means that mixing is energetically favourable. When the interaction parameter χ falls between zero and two, mixing is energetically unfavourable, but the gain in entropy on mixing is always large enough to offset this.

In order to obtain useful predictions of phase behaviour from this model which can be compared to experiment, it is necessary to have some prediction for the way the interaction parameter χ depends on temperature. In the simplest interpretation of the model, the interactions $\epsilon_{AA}, \epsilon_{BB}, \epsilon_{AB}$ were regarded as being purely energetic in character; in this case they are temperature independent and the interaction parameter must vary as $1/T$. In this simple case, we find that a mixture will be phase separated at low temperature, but will form a single phase at higher temperatures. This is to be expected, as a single phase is more disordered than a phase-separated system.

[1] It must be stressed that this relationship is not in general the definition of the coexistence curve; it applies only in this situation where the free energy is a symmetrical function of ϕ. In general, the coexistence curve must be determined from the condition that the chemical potentials in each phase are equal. This condition is met when one draws a single straight line which is tangent to the free energy curve at the two coexisting compositions—the **double-tangent** construction.

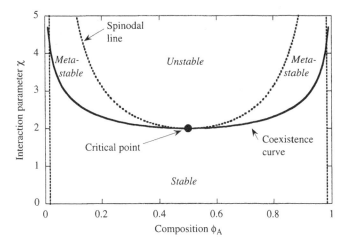

Fig. 3.5 The phase diagram of a liquid mixture whose free energy of mixing is described by the regular solution model (eqn 3.6).

However, the interactions that give rise to χ can include more specific interactions such as hydrogen bonds as well as van der Waals forces, and it is often found that these cannot be interpreted simply in terms of energy changes; they may also have an entropic component. In these cases the temperature dependence of χ may be more complicated.

3.2.1 Interfaces between phases and interfacial tension

A pair of immiscible liquids, like oil and water, are separated by an interface, and it is a familiar observation that such interfaces possess a certain property of **tension**, just as the surface of a liquid does. A droplet of oil in water will adopt a spherical shape, to minimise the area of interface between the two liquids.

Interfacial tension is equivalent to an interfacial free energy, as can be seen from the thought experiment illustrated in Fig. 3.6. Here a barrier defines the area of an interface between two immiscible liquids; to keep the barrier in position a force F must be applied to it to counteract the force due to the interfacial tension γ. Thus $F = \gamma L$. If the barrier is moved a distance x, the total work done is given by $Fx = \gamma L x$. In doing this we have created an area Lx of new interface, so we can identify the interfacial tension γ with the energy per unit area involved in creating the new interface.

An important subtlety arises when we consider what kind of energy this corresponds to. We would normally do such an experiment in conditions of constant temperature, rather than in conditions in which the experiment is isolated from the rest of the world—that is to say, we would use **isothermal** rather than **adiabatic** conditions. Thus during the process of making more interface heat can flow in or out of the system in order to keep the system at constant temperature. Thus the interfacial tension is an interfacial **free energy** rather than simply an internal energy; we expect there to be some entropy associated with the interface.

Where does this interfacial free energy come from at a microscopic level? The unfavourable energy must be associated with contacts between the A and B molecules. If the interface were absolutely sharp on a molecular scale, then we could use the kind of model introduced in the last section to deduce

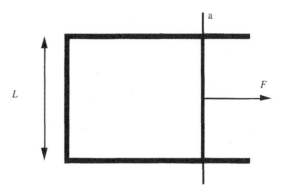

Fig. 3.6 A thought experiment to illustrate the relation between interfacial tension and interfacial energy. A movable barrier (a) defines the area of an interface between two immiscible fluid phases (shown shaded). In order to maintain the position of the barrier a force F must be imposed to counteract the force due to the interfacial tension γ; at equilibrium $F = \gamma L$.

a relationship between γ_{sharp} and the pairwise interaction energies ϵ_{AA}, ϵ_{BB}, and ϵ_{AB}. In fact, it is easy to show in this picture that

$$\gamma_{\mathrm{sharp}} = \frac{1}{2v^{2/3}}(2\epsilon_{AB} - \epsilon_{AA} - \epsilon_{BB}) = \frac{\chi k_B T}{z v^{2/3}}, \tag{3.10}$$

where v is a molecular volume. In reality the interfacial tension will always be less than this, because the interface will not be absolutely sharp. The roughening and diffuseness of the interface that result from the Brownian motion of molecules at the interface mean that the interface will have an excess entropy associated with it, and this excess entropy leads to a reduction of the free energy of the interface at finite temperatures. Thus we expect the interfacial tension to decrease significantly as the temperature is increased. As we approach a critical point, we would expect the distinction between the two phases to vanish, and the interface to disappear. Thus as we approach the critical point the interfacial tension must approach zero.

3.3 Liquid–liquid unmixing—kinetics of phase separation

3.3.1 Two mechanisms of phase separation

In the last section we introduced the distinction between compositions that are unstable and compositions that are metastable. This distinction is important when we consider the mechanism by which phase separation takes place. In the unstable part of the phase diagram, as the name suggests, phase separation takes place by a **continuous** change in composition, by a process in which the concentration fluctuations that are present in any mixture at thermal equilibrium are amplified. The process by which this happens is known as **spinodal decomposition**. In contrast, when a mixture is in the metastable region of the phase diagram, it is not possible for the mixture to phase-separate by a process in which the composition in a region changes continuously; in a pure mixture a relatively large composition fluctuation must take place, with a corresponding high energy cost. This relatively large **nucleus** can then grow in size. This process is known as **homogeneous nucleation**. In a real system, it is usually found that some impurity particles are present, on which the new phase may be nucleated with lower activation energy than for homogeneous nucleation. This situation is known as **heterogeneous nucleation**.

3.3.2 Spinodal decomposition

If we are within the spinodal line, any small local change in composition lowers the free energy; the system is unstable, and any small fluctuation in composition is amplified. This process is known as spinodal decomposition. In spinodal decomposition, material flows from regions of low concentration to regions of high concentration. This is a reversal of the normal situation, in which we expect material to diffuse from regions of high concentration to regions of low concentration. This process is sometimes known as uphill diffusion.

Why, in spinodal decomposition, do we get this reversal in the direction of diffusion? We are used to the idea that diffusion acts to make the concentration uniform. This is usually true, but in general the fundamental quantity that must be uniform at equilibrium is not the concentration, but the **chemical potential**. Material will diffuse down the gradient of chemical potential, from regions of high chemical potential to regions of low chemical potential. The chemical potential is related to the first derivative of the free energy with respect to concentration; this means that if the second derivative of free energy with respect to concentration is positive then regions of high concentration have high chemical potential and diffusion is in the normal, downhill direction. But inside the spinodal region the second derivative of free energy with respect to concentration is negative, the chemical potential gradient has the opposite sign to the concentration gradient, and material flows from regions of low concentration to regions of high concentration—uphill diffusion.

Thus any fluctuation in composition will grow, ultimately leading to a phase-separated domain at the coexisting composition. However, not all concentration fluctuations grow at the same rate. The growth of long-wavelength fluctuations involves the diffusion of molecules over long distances, which is relatively slow; on the other hand if very short-wavelength fluctuations were to grow they would create a large amount of interface, which itself would cost too much energy. There is thus an optimal size of concentration fluctuation which grows the fastest. Patterns created by spinodal decomposition are random, but have a characteristic length scale (see Fig. 3.7).

How can we make this picture more quantitative? To do this, we need to be able to take into account the energy associated with the interfaces between the incipient domains as they grow. These interfaces are not yet sharp, however, so we cannot simply associate with them a certain energy per unit area. Instead, we must recognise that the local free energy density of a mixture must depend not just on the local composition ϕ, but also on the gradient of composition. The simplest way of expressing this is to write the free energy density with a term proportional to the square of the concentration gradient. We write the total free energy of the system F as a **functional** of the volume fraction $\phi(x)$, which itself is a function of the position coordinate x (for simplicity, we work

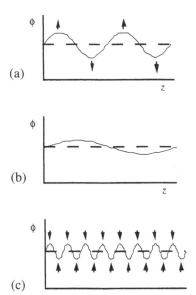

Fig. 3.7 Why concentration fluctuations of one particular intermediate length scale grow fastest in spinodal decomposition. In (b), a long-wavelength fluctuation grows relatively slowly, as the distances material has to diffuse from troughs to peaks are relatively large. In (c), too much new interface is created, with a correspondingly large free energy penalty; these fluctuations do not grow at all. An intermediate length scale (a) grows the fastest and dominates the pattern of phase separation.

in one dimension). Thus we can write

$$F = A \int \left[f_0(\phi) + \kappa \left(\frac{d\phi}{dx} \right)^2 \right] dx, \tag{3.11}$$

where $f_0(\phi)$ is the free energy per unit volume for a uniform mixture of composition ϕ, and κ is the **gradient energy coefficient**.

This type of theory has a long history in a variety of different physical contexts; it was introduced by van der Waals for the study of the liquid–gas interface, was used by Landau and Ginsburg for the study of (amongst other things) magnetic domain walls, and by Cahn and Hilliard in the context of interfaces and spinodal decomposition in metals. The theory is entirely phenomenological; this is a strength in that it can be expected to apply to a wide variety of different situations. A weakness is that the value of the gradient energy coefficient κ is not specified without further theoretical work appropriate to the situation being considered. One should note that κ has the dimensions of a length squared; the length in question must be related to the range of the intermolecular forces involved, and may be expected to be of the same order as intermolecular distances in simple fluids.

How does a system whose free energy can be defined by an expression of this kind evolve in time? Systems with non-uniform compositions generally evolve according to **diffusion equations**, and this system is no different, though as we shall see, the form of the free energy implied by eqn 3.11 means that the diffusion equation is somewhat modified. Let us first recall that we normally define a diffusion equation by Fick's first law, which relates the flux J of material to the concentration gradient via the diffusion coefficient D as

$$J = -D \frac{d\phi}{dx}, \tag{3.12}$$

where we once again work in one dimension for the sake of simplicity. Local conservation of material is expressed by the equation of continuity,

$$\frac{d\phi}{dt} = -\frac{dJ}{dx}, \tag{3.13}$$

and combining this with eqn 3.12 yields Fick's second law of diffusion,

$$\frac{\partial \phi}{\partial t} = D \frac{\partial^2 \phi}{\partial x^2}. \tag{3.14}$$

The fundamental difficulty with this approach is that it assumes that diffusion is driven by gradients in **composition**. However, statistical mechanics tells us that at equilibrium it is not necessarily the composition that must be constant everywhere, but the **chemical potential**. For small departures from equilibrium, we should expect the flux of material to be proportional, not to the gradient of concentration, but to the gradient of chemical potential. For a binary mixture in which the total density is constrained to be constant, we expect as the counterpart of eqn 3.12 an equation of the form

$$J_A = -M \frac{d}{dx} (\mu_A - \mu_B) \tag{3.15}$$

for the flux of species A, J_A. Here M is a positive transport coefficient, the **Onsager coefficient**, and μ_A and μ_B are the chemical potentials of A and B. The so-called **exchange chemical potential**, $\mu = \mu_A - \mu_B$, represents the change in free energy when one A molecule is removed and replaced by a molecule of B; this can be written in terms of a derivative with respect to concentration of the free energy density:

$$\mu = \frac{d}{d\phi}\left[f_0(\phi) + \kappa \left(\frac{d\phi}{dx}\right)^2 \right]. \tag{3.16}$$

When we calculate the free energy, we integrate over the expression in square brackets with respect to distance; one can show, using integration by parts, that this expression can be rewritten as

$$\mu = \frac{d f_0}{d\phi} + 2\kappa \frac{d^2\phi}{dx^2}. \tag{3.17}$$

Substituting this into 3.15 we find for the flux

$$-J_A = M f_0'' \frac{\partial \phi}{\partial x} + 2M\kappa \frac{\partial^3 \phi}{\partial x^3}. \tag{3.18}$$

Here we have written $f_0'' = d^2 f_0/d\phi^2$. This quantity is certainly a function of the concentration, and M and κ could in principle also be concentration dependent. Nonetheless, in order to keep the problem soluble we *assume* that they are independent of concentration; this **linearisation** of the equation means that our solutions will be valid only in the limit of small deviations from some uniform starting concentration ϕ_0. Now combining eqn 3.18 with the continuity equation 3.13 we obtain the modified diffusion equation whose solutions give us the time evolution of a phase-separating mixture—the **Cahn–Hilliard equation**:

$$\frac{\partial \phi}{\partial t} = M f_0'' \frac{\partial^2 \phi}{\partial x^2} + 2M\kappa \frac{\partial^4 \phi}{\partial x^4}. \tag{3.19}$$

One can see that in the absence of the gradient term, which is proportional to κ, this is simply the diffusion equation, Fick's second law, where we can identify an effective diffusion coefficient D_{eff} as

$$D_{\text{eff}} = M f_0''. \tag{3.20}$$

The important point about this effective diffusion coefficient is that while M always must be positive, f_0'' can be either positive or negative. In particular, since the definition of the spinodal line is that there $d^2 f/d\phi^2 = 0$, inside the spinodal line the effective diffusion coefficient is *negative*. Material diffuses from regions of low concentration to regions of high concentration. What is the solution of the Cahn–Hilliard equation? One can verify by direct substitution that one set of solutions is given by

$$\phi(x, t) = \phi_0 + A \cos(qx) \exp\left[-D_{\text{eff}} q^2 \left(1 + \frac{2\kappa q^2}{f_0''}\right) t \right]. \tag{3.21}$$

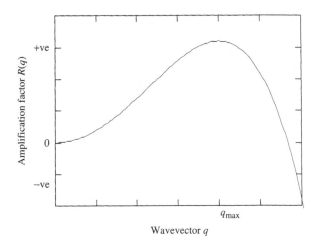

Fig. 3.8 The amplification factor for spinodal decomposition in the linear theory.

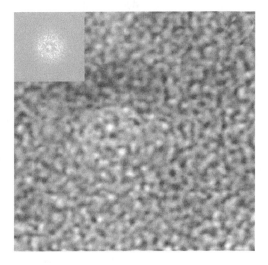

Fig. 3.9 Pattern formed by spinodal decomposition of a mixture of two polymers in a common solvent. Confocal micrograph of a mixture of polystyrene, polybutadiene, and toluene after quenching for 82 seconds. Picture courtesy of B. Jung.

So inside the spinodal line, any composition fluctuation with wavevector q grows exponentially, with a q-dependent **amplification factor** $R(q)$ given by

$$R(q) = -D_{\text{eff}}q^2 \left(1 + \frac{2\kappa q^2}{f_0''}\right). \tag{3.22}$$

This function is plotted in Fig. 3.8; it has a maximum at a wavevector q_{max}, which defines the fastest-growing wavevector. This sets the length scale on which spinodal decomposition occurs.

How do we recognise spinodal decomposition experimentally? If we look at a sample undergoing spinodal decomposition with a microscope we would expect to see a random pattern (because the pattern of domains results from the amplification of random fluctuations in composition), but with a characteristic length corresponding to the wavelength of the fastest-growing fluctuation $2\pi/q_{\text{max}}$. An example of the type of pattern is shown in Fig. 3.9.

This kind of pattern is random, but it is not without structure; the existence of a characteristic length is most clearly demonstrated by taking a Fourier transform (see inset to Fig. 3.9). This shows a ring, and a radial average of the

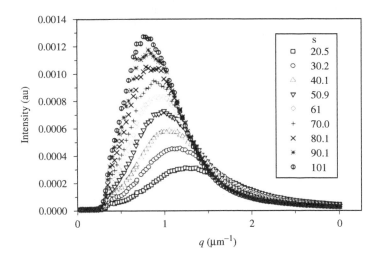

Fig. 3.10 Light-scattering curves from a polymer mixture quenched into the unstable region of the phase diagram, showing the maximum in intensity at q_{max} characteristic of spinodal decomposition. q_{max} moves to smaller values at longer annealing times, owing to non-linear coarsening effects. Graph courtesy of B. Jung.

pattern would show a peak at a wavevector corresponding to the characteristic length scale. Rather than taking Fourier transforms of individual micrographs, it is easier practically to do a scattering experiment on the sample, which yields the Fourier transform directly. The resulting data is illustrated in Fig. 3.10. This shows the evolution of scattering with time; initially a peak appears at a value of wavevector which should correspond to the fastest-growing wavevector as predicted by the linear theory. This peak grows in intensity as the phase separation proceeds. As the peak grows, it also moves into smaller wavevectors. This corresponds to the characteristic length scale growing with time. This is not predicted by the simple linear theory; in mathematical terms it is a consequence of the non-linearity of eqn 3.19. Physically, the system evolves larger domains in order to reduce the total interfacial energy of the system. This growth in domain size is known as **coarsening**, and we shall discuss it in more detail in the next section.

Analysis of such curves can be carried out in the framework of the Cahn–Hilliard theory; this analysis should yield both the effective diffusion coefficient and the value of the second derivative of free energy with respect to concentration.

3.3.3 Nucleation

As we discussed above, a mixture quenched into the metastable part of the phase diagram is stable with respect to small changes in concentration, and so the continuous process of spinodal decomposition is not available to it. Instead, a drop of material at the other coexisting composition must somehow come into being despite this causing an increase in free energy, and then subsequently grow until the free energy starts to decrease again. In other words, a particle at the other coexisting composition must be **nucleated** by a thermal fluctuation. Let us ask what energy is required to nucleate a particle of size r. There will be a negative contribution to the free energy which is proportional to the **volume** of the droplet; this is because the mixture is initially globally unstable so that if it does succeed in phase separating this will lower the free energy by an amount ΔF_v per unit volume. However, to make a droplet one needs to make an interface; there is an interfacial energy γ and this leads to a positive contribution to the free energy proportional to the **surface area** of the droplet. Thus we can write the net change in free energy on nucleating a droplet of size r, $\Delta F(r)$, as

$$\Delta F(r) = \frac{4}{3}\pi r^3 \Delta F_v + 4\pi r^2 \gamma. \tag{3.23}$$

This energy has a maximum value for a critical size r^*; droplets of a size below r^* are unstable, in that once formed the energy of the system increases if they grow any further. Droplets with a size above r^* are stable, and when they grow the free energy of the system is lowered. The energy required to form this critical nucleus ΔF^* may be easily calculated; it is given by

$$\Delta F^* = \frac{16\pi \gamma^3}{3\Delta F_v^2}. \tag{3.24}$$

Nucleation is an **activated** process; it can only occur if a fluctuation occurs increasing the local free energy by an amount ΔF^*. The probability that this happens is given by a Boltzmann factor; thus the rate of nucleation of droplets is proportional to $\exp(-\Delta F^*/k_B T)$.

The barrier for nucleation is usually greatly reduced by the presence of dust particles, or indeed the walls of the vessel. This kind of **heterogeneous** nucleation is in practice much more common than the **homogeneous** nucleation process described here.

3.3.4   Growth in the late stages of phase separation

Once phase separation has proceeded to the extent that domains of one phase have begun to form, whether by spinodal decomposition or homogeneous nucleation, these domains subsequently grow. The driving force for growth is the reduction in interfacial energy that occurs as the domains get larger. The laws describing the subsequent growth are complicated and poorly understood, except in certain limiting cases.

One such limiting case is when there are small numbers of minority phase particles, with a distribution of particle sizes. The effect of curvature is such that the local solubility of the minority component is higher near the smaller particles than the larger ones; the smaller particles dissolve and the material is absorbed by the larger ones. This process, by which larger particles grow at the expense of the smaller ones, is known as **Ostwald ripening**. One can show (see e.g. Lifshitz and Pitvaevski (1981)) that the average drop size grows as the one-third power of time—the **Lifshitz–Slyozov** law.

A more general approach considers the way in which the length scales characterising domain sizes and interface widths change with time (see Fig. 3.11). In spinodal decomposition, the incipient domains arise as a result of the amplification of sinusoidal composition fluctuations. The domain size $R(t)$ and the width of the interface between domains $w(t)$, inasmuch as it makes sense to define the latter quantity in a system which in a sense is all interface, are essentially identical, and in the **early** stage of phase separation do not change with time. Towards the end of the early stage, the compositions begin to approach the equilibrium coexisting compositions. At this point the average domain size $R(t)$ begins to grow, while the interface width $w(t)$ must start to

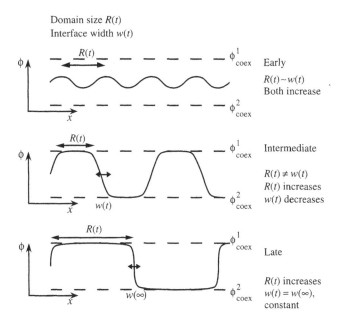

Domain size $R(t)$
Interface width $w(t)$

Early
$R(t) \sim w(t)$
Both increase

Intermediate
$R(t) \neq w(t)$
$R(t)$ increases
$w(t)$ decreases

Late
$R(t)$ increases
$w(t) = w(\infty)$,
constant

Fig. 3.11 Sketches of composition profiles in a phase-separating system, illustrating how the length scales characterising the average domain size $R(t)$ and interfacial width $w(t)$ change with time.

shrink as the initially rather diffuse incipient domains become much more well defined. This defines a rather complex intermediate stage of phase separation. Finally, the interfaces have sharpened up to their equilibrium values, while the average domain size $R(t)$ continues to grow. It is in this late stage of phase separation that some simplicity returns.

In the late stage of phase separation, the **dynamical scaling hypothesis** suggests that there is only one important length scale characterising all aspects of the structure, and that this length scale is the average domain size $R(t)$. The implication of this is that if we characterise the pattern by some statistical measure such as the two-point correlation function $G(r, t)$, then we can write this as a universal function of a single reduced variable x, so

$$G(r, t) = G(x) \quad \text{where} \quad x = \frac{r}{R(t)}. \tag{3.25}$$

The physical meaning of this is that statistically the pattern we see at any later time is simply a magnification of the pattern at an earlier time. Thus we have reduced the complicated problem of determining the spatial characteristics of the domain pattern at any arbitrary time to the two simpler problems of determining the single scaled pattern $G(x)$, valid for all times in the late stage, and of finding out the way the average domain size $R(t)$ varies with time.

Obtaining a general theory to predict the scaling function $G(x)$ seems to be difficult. However, the behaviour of the average domain size $R(t)$ is more tractable. It turns out that, provided that the mechanism of transport of matter is by diffusion (rather than, say, bulk flow of a liquid), then the Lifshitz–Slyozov law, that $R(t) \sim t^{1/3}$, is always valid, not just for the case of small concentrations of spherical minority phase particles for which an exact solution is available. That this is so can be seen by a scaling argument relying on the assertion that the only relevant length scale in the problem is $R(t)$.

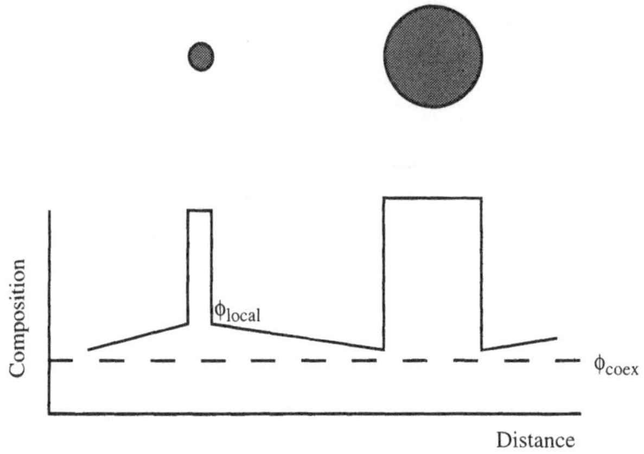

Fig. 3.12 Schematic composition profile near two drops. The local composition ϕ_{local} near the boundary of the smaller drop is larger than the bulk coexisting composition ϕ_{coex}; this leads to a concentration gradient which drives diffusion of material from small drops to large drops.

The situation is sketched in Fig. 3.12. The key point is that the composition near smaller drops is slightly higher than the composition near larger drops, and this leads to a composition gradient down which material flows, to feed the growth of the large drop at the expense of the small drop. It is the curvature of the surface of the small drop that leads to the local composition being higher than the bulk composition. This is the same phenomenon as the **Gibbs pressure** in soap bubbles, and similarly we find that the increase in local composition near a droplet of size r is given by

$$\phi_{\text{local}} - \phi_{\text{coex}} \sim \frac{\gamma}{r}. \tag{3.26}$$

The argument then proceeds as follows. When the length scale $\sim R(t)$, this leads to curvatures $\sim R(t)$. This leads to concentration differences of typical magnitude $\sim \gamma/R(t)$, which in turn leads to concentration gradients $\sim \gamma/R(t)^2$. We can estimate the resulting diffusion fluxes $J \sim D\gamma/R(t)^2$. It is this flux of material J that leads to growth of the domains:

$$\frac{\mathrm{d}R(t)}{\mathrm{d}t} \sim J, \tag{3.27}$$

so we have

$$\frac{\mathrm{d}R(t)}{\mathrm{d}t} \sim \frac{D\gamma}{R(t)^2}. \tag{3.28}$$

On integration, this relation gives us the growth law

$$R(t) \sim (D\gamma t)^{1/3}. \tag{3.29}$$

This relation is illustrated with some experimental data in Fig. 3.13. This data shows the way the peak wavevector Q_{m} varies with reduced time τ in a phase-separating mixture of polymers in a common solvent for a variety of different temperatures. Of course, $Q_{\text{m}} \sim R(t)^{-1}$, so we expect $Q_{\text{m}} \sim \tau^{-1/3}$, which is close to what is experimentally observed.

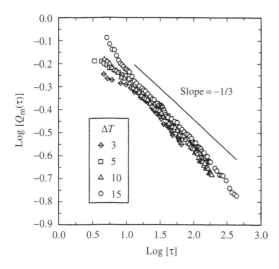

Fig. 3.13 Peak wavevector Q_m plotted against reduced time τ as measured in light-scattering experiments on a phase-separating mixture of two polymers in a common solvent at various temperature increments below the critical temperature ΔT. Graph courtesy of B. Jung.

3.4 The liquid–solid transition—freezing and melting

The transition between liquid and solid is familiar to everyone, and important both for normal matter and for soft matter. When a liquid freezes, it goes from a state in which there is no long-range order to a state in which every molecule is associated with a definite lattice position, whose position relative to every other lattice position is well defined. Associated with this long-ranged order is the rigidity that we associate with a solid. The theory of this transition is more difficult than for the liquid–liquid unmixing transition, even at mean field level. This is because, whereas the progress of the liquid–liquid transition can be described in terms of a single **order parameter**, the solid–liquid transition needs to be characterised in terms of an infinite number of such parameters. If we imagine the density, averaged over time, along a line in a liquid, we would find a single, constant, value. The onset of solid-like order in the material would be marked by peaks in the density growing at the lattice positions; to describe the perfection of the solid-like order one would need to specify the values of all the Fourier components of the density in the lattice. It is these values that constitute the order parameters for the liquid–solid transition. A mean field theory of the transition can be constructed in terms of these parameters, but this is beyond the scope of this book—the interested reader is referred to Chaikin and Lubensky (1995). Instead, here, we introduce the phenomenology of the liquid–solid transition and use this to discuss various aspects of the kinetics of freezing.

The liquid–solid transition is a **first-order phase transition**; this means that at the transition the state of order changes discontinuously, and thermo-dynamic quantities that are derivatives of a free energy with respect to other thermodynamic variables are discontinuous at the transition. This is illustrated in Fig. 3.14. When a liquid turns into a solid at the melting temperature a latent

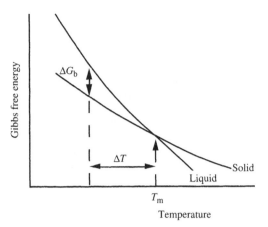

Fig. 3.14 Schematic diagram of the variation of Gibbs free energy with temperature near a melting point at T_m. At a degree of undercooling ΔT the change in free energy per unit volume on going from the liquid state to the solid state is ΔG_b.

heat ΔH_m is released; this is related to the change in entropy on freezing ΔS_m by

$$\Delta S_m = \frac{\Delta H_m}{T_m}. \tag{3.30}$$

From an inspection of Fig. 3.14 we can argue that a liquid held exactly at its melting temperature, if it is free from impurities and in a large enough container to be unaffected by the walls, will never freeze. This is because the creation of a crystal must **cost** free energy, and to do so will create an interface between the solid and liquid that will have a certain interfacial energy γ_{sl} associated with it. But the change in free energy per unit volume going from liquid to solid is zero exactly at the melting point, because this is defined as the temperature at which the free energies of liquid and solid are the same. Thus in order to initiate freezing in the absence of impurities, one must **undercool** a liquid below its melting point. Freezing is then initiated by an **activated** process of nucleation.

3.4.1   Kinetics of the liquid–solid transition—homogeneous nucleation

If crystallisation in an undercooled melt is initiated by the spontaneous appearance of a crystal nucleus of radius r (for the moment we assume the crystal to be spherical), then we can write the change in free energy $\Delta G(r)$ as the sum of a term proportional to the surface area of the crystal, representing the contribution of the solid–liquid interface, with interfacial energy γ_{sl}, and a term proportional to the volume, representing the change in Gibbs free energy on going from liquid to solid. This argument is very similar to that presented in Section 3.3.3. Thus we have

$$\Delta G(r) = \frac{4}{3}\pi r^3 \Delta G_b + 4\pi r^2 \gamma_{sl}. \tag{3.31}$$

By inspection of Fig. 3.14, we can rewrite this in terms of the latent heat of melting. Writing the entropy change on melting in terms of the latent heat, we have

$$\Delta S_m = \left(\frac{\partial G_s}{\partial T}\right)_P - \left(\frac{\partial G_l}{\partial T}\right)_P = \frac{\Delta H_m}{T_m}, \tag{3.32}$$

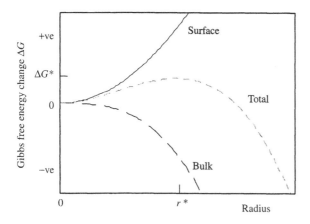

Fig. 3.15 Schematic diagram of the change in Gibbs free energy when a crystal of radius r is nucleated in a melt cooled by ΔT below its melting point.

where the subscripts s and l refer to the solid and liquid respectively. If we can assume that the undercooling ΔT is small enough that over that temperature range the partial derivatives are approximately constant, we can write for the free energy change per unit volume when a melt undercooled by ΔT freezes, ΔG_{b},

$$\Delta G_{\mathrm{b}} = -\frac{\Delta H_{\mathrm{m}}}{T_{\mathrm{m}}} \Delta T. \tag{3.33}$$

Using this in eqn 3.31 we have for the free energy change on nucleating a crystal radius r

$$\Delta G(r) = -\frac{4}{3}\pi r^3 \frac{\Delta H_{\mathrm{m}}}{T_{\mathrm{m}}} \Delta T + 4\pi r^2 \gamma_{\mathrm{sl}}. \tag{3.34}$$

This function is sketched in Fig. 3.15. The free energy change has a maximum at a critical crystal radius r^*, given by

$$r^* = \frac{2\gamma_{\mathrm{sl}} T_{\mathrm{m}}}{\Delta H_{\mathrm{m}} \Delta T}. \tag{3.35}$$

Crystals bigger than r^* can continue to grow, since by doing so they will lower the free energy. Crystals less than r^* are unstable and will remelt. There is a free energy barrier ΔG^* associated with the critical nucleus size; this is the barrier that must be overcome by thermal fluctuation in order for a viable crystal to be nucleated. This energy barrier is given by

$$\Delta G^* = \frac{16\pi}{3}\gamma_{\mathrm{sl}}^3 \left(\frac{T_{\mathrm{m}}}{\Delta H_{\mathrm{m}}}\right)^2 \frac{1}{\Delta T^2}. \tag{3.36}$$

The probability of a crystal being nucleated is proportional to

$$\exp(-\Delta G^*/k_{\mathrm{B}}T).$$

This is an extraordinarily strong function of temperature; using typical values this equation predicts that one should only be able to observe significant rates of nucleation tens of degrees below the melting point.

Our experience of liquids freezing is, of course, that one does observe growth of crystals in the liquid for undercoolings of only a few degrees, if that.

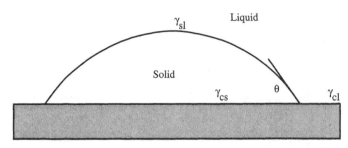

Fig. 3.16 A crystal nucleus heterogeneously nucleated against a solid surface which acts as a catalyst for crystallisation by lowering the activation energy for nucleation. The contact angle θ is related to the interfacial tensions γ_{cs}, γ_{cl}, and γ_{ls} between catalyst and solid, catalyst and liquid, and liquid and solid respectively.

The reason for this is that the theory sketched above, describing homogeneous nucleation, is relevant only for pure liquids. In reality liquids contain particles of dust or other contaminants, and these provide nucleation sites for crystal growth with much lower activation barriers than for homogeneous nucleation. That this is so can be seen in the next section.

3.4.2 Kinetics of the liquid–solid transition—heterogeneous nucleation

If a pre-existing surface of a different solid is present in an undercooled melt, this usually lowers the activation energy for nucleation of a new crystal, often very substantially, and thus acts as a catalyst for crystallisation. Such a surface may be the surface of a piece of dust or other contaminant, or even the walls of the container that the liquid is held in.

Figure 3.16 illustrates the spherical cap of solid that will be formed against a planar catalyst surface. The relationship between the contact angle θ and the interfacial tensions γ_{cs}, γ_{cl}, and γ_{ls} between catalyst and solid, catalyst and liquid, and liquid and solid respectively, is given by Young's equation, which may be derived by balancing the forces at the contact line:

$$\gamma_{sl} \cos \theta = \gamma_{cl} - \gamma_{cs}. \tag{3.37}$$

If the radius of curvature of the spherical cap is r, then its volume V is given by

$$V = \frac{1}{3}\pi r^3 (1 - \cos \theta)^2 (2 + \cos \theta). \tag{3.38}$$

The area of the solid/liquid interface S_{sl} is

$$S_{sl} = 2\pi r^2 (1 - \cos \theta) \tag{3.39}$$

and the area of the catalyst/solid interface S_{cs} is

$$S_{cs} = \pi r^2 \sin^2 \theta. \tag{3.40}$$

If we now repeat the argument of the last section to find the free energy barrier ΔG^* for nucleation with interface and volume contributions

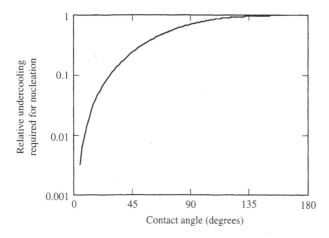

Fig. 3.17 The rate of nucleation for heterogeneous nucleation of a crystal from an undercooled melt.

appropriate for the spherical cap we find

$$\Delta G^* = \frac{16\pi}{3} \gamma_{sl}^3 \left(\frac{T_m}{\Delta H_m} \right)^2 \frac{1}{\Delta T^2} \frac{(1 - \cos\theta)^2 (2 + \cos\theta)}{4}. \quad (3.41)$$

For contact angles less than 90° this geometrical factor dramatically decreases the degree of undercooling required to achieve a measurable nucleation rate. This is illustrated in Fig. 3.17, which plots the degree of undercooling required to achieve a given nucleation rate relative to the undercooling required for homogeneous nucleation. This diagram makes clear how effective some solid surfaces can be at catalysing crystallisation from the melt.

The most effective catalyst for the formation of a solid from an undercooled melt is a seed crystal of the solid. This reduces the activation barrier for crystallisation to zero. However, even if no nucleation step is required the process of the growth of a crystal front into a melt itself has interesting and complex features, which we discuss in the next section.

3.4.3 Solidification—stability of a growing solidification front

A solidification front moving into an undercooled melt is unstable. This instability can lead, according to conditions, into cellular or dendritic growth. These dendritic patterns are familiar to anyone who has watched ice form on the inside of a window in cold weather, and are responsible for the spectacular shapes of the ice crystals in snow. Although many details of the way in which dendritic growth leads to this kind of pattern are not yet fully worked out, the basic mechanism underlying the instability of a growing crystal front was worked out by Mullins and Sekerka.

The instability arises from the fact that when liquid crystallises, it releases latent heat; before further crystallisation can occur this heat must diffuse away. Imagine a solid front moving into an undercooled liquid. Figure 3.18 shows a sketch of the temperature distribution, in a situation where (for simplicity) the temperature in the solid takes the value at its melting point (this corresponds to the case where the thermal conductivity of the liquid is greater than that of the

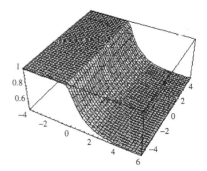

Fig. 3.18 The temperature distribution at a planar interface between a crystal and its undercooled melt. The rate of growth of the interface is controlled by the rate at which the latent heat can diffuse away down the temperature gradient.

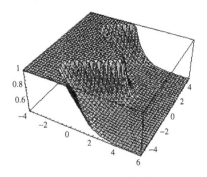

Fig. 3.19 The temperature distribution at an interface between a crystal and its undercooled melt that has become wavy. The rate of growth at the peak of the wave is faster than at the trough, because the temperature gradient is steeper there and the latent heat can diffuse away faster.

solid). The rate at which the solid front can grow is limited by the rate at which heat can diffuse away down the temperature gradient.

This picture of a planar front growing into its undercooled melt is not necessarily stable. If a fluctuation causes the solid/liquid interface to become wavy (see Fig. 3.19), the growth rate at the peak of the wave is faster than at the trough, because at the peak the temperature gradient is greater there and the latent heat can diffuse away faster. Thus this kind of fluctuation can be unstable and will be magnified.

Not all fluctuations will be equally unstable. We might imagine that there is a wavelength of fluctuation which will grow fastest; short-wavelength fluctuations will lead to an increase in the interfacial area between the solid and the melt, and thus will be suppressed, while very long-wavelength fluctuations will not produce such large increases in temperature gradient. This kind of argument can be made quantitative by considering the stability of the interface against small perturbations. This linear theory predicts that for a front growing into a liquid with velocity v, fluctuations with wavevector k grow with an amplification factor $\omega(k)$ given by

$$\omega(k) = kv - (1 + \beta)D_T d_0 k^3, \tag{3.42}$$

where D_T is the thermal diffusivity, β is the ratio of the product of thermal diffusivity and heat capacity in the solid and liquid phases (which will be of order unity), and d_0 is a capillary length given by

$$d_0 = \frac{\gamma_{sl} T_m c_p}{\Delta H_m}, \tag{3.43}$$

with c_p the specific heat capacity of the liquid. These expressions predict that all perturbations with a wavevector shorter than k_s, given by

$$k_s = \sqrt{\frac{v}{(1 + \beta)D_T d_0}}, \tag{3.44}$$

are unstable. Thus long-wavelength perturbations are unstable, with the minimum wavelength of order the geometric mean of the diffusion length and the capillary length.

Here we have discussed the problem of a solidification front moving into a pure undercooled liquid. In many real situations, the melt contains some impurities. In this case, for the solidification front to advance the impurities must diffuse away. This leads to an instability with very similar physics, resulting in dendritic or cellular growth with the impurities forced to the gaps between the dendrites.

Further reading

A comprehensive and much more sophisticated treatment of the statistical mechanics of phase transitions in condensed matter can be found in Chaikin and Lubensky (1995). A review of the theory of spinodal decomposition, showing how the approximations of the simple linear theory described here may be lifted, is given by Binder (1991).

The theory of the instability underlying dendritic growth was first given by Mullins and Sekerka (1964). A good non-technical account of dendritic growth and its relationship to other growth instabilities can be found in Ball (1999).

Exercises

(3.1) The phase behaviour of a certain liquid mixture can be described by the regular solution model, with the interaction parameter being given by $\chi = 600/T$, where T is the temperature in kelvin. Calculate the following quantities:

  a) The temperature at the critical point.
  b) The volume fractions of the coexisting compositions at 273 K.
  c) The volume fractions on the spinodal line at 273 K.

(3.2) a) Use regular solution theory to derive an approximate expression, valid in the limit of a large, positive value of the interaction parameter χ, for the limit of solubility of one immiscible liquid in another.
  b) Comment on your result.

(3.3) The value of χ between water and linear hydrocarbons may be taken to be given by the formula $\chi = 3.04 + 1.37n_C$, where n_C is the number of carbon atoms in the hydrocarbon. Use the formula derived in the last question to compare the limiting solubilities in water of hexane (C_6H_{14}), octane (C_8H_{18}), and decane ($C_{10}H_{22}$).

(3.4) Estimate the interfacial tension between octane and water. You may take the interaction parameter between octane and water as $\chi = 14.0$, and the molecular volume as $2.36 \times 10^{-29} \, m^3$.

(3.5) A light-scattering experiment is carried out on a phase-separating polymer mixture. Values of intensity are recorded as a function of time at a variety of scattering angles q. The data is shown below.

$Q(\mu m^{-1})$	0.970	1.210	1.460	1.860	2.780	3.000
Time (s)	$\ln(I(Q,t))$					
0.983	−11.920	−12.170	−12.100	−11.900	−10.910	−11.280
1.966	−11.570	−11.610	−11.260	−11.010	−10.430	−10.560
2.950	−11.500	−11.050	−10.700	−10.470	−10.150	−10.190
3.575	−11.210	−10.630	−10.320	−10.100	−10.000	−10.000
4.916	−10.880	−10.250	−9.997	−9.807	−9.898	−9.901
5.899	−10.590	−9.997	−9.693	−9.579	−9.854	−9.872
6.972	−10.390	−9.763	−9.482	−9.455	−9.832	−9.882
7.866	−10.140	−9.529	−9.295	−9.269	−9.810	−9.872
8.849	−9.987	−9.389	−9.178	−9.207	−9.832	−9.941
9.654	−9.755	−9.225	−9.015	−9.083	−9.825	−9.941
10.820	−9.600	−9.061	−8.898	−9.041	−9.847	−9.951
11.710	−9.445	−8.968	−8.827	−8.979	−9.861	−9.990
12.780	−9.290	−8.851	−8.757	−8.938	−9.876	−10.020
13.680	−9.155	−8.734	−8.687	−8.876	−9.876	−10.030
14.750	−9.077	−8.640	−8.594	−8.834	−9.883	−10.050

  a) Plot the scattered intensity as a function of time for $Q = 1.21 \, \mu m^{-1}$. Explain the shape of the curve.
  b) Use the data to extract values for the amplification factor, $R(Q)$, as a function of scattering wavevector Q.

  c) How, according to Cahn–Hilliard theory, do you expect $R(Q)$ to vary with Q? Plot your values of $R(Q)$ in a way that tests this theory and comment on the degree of agreement or otherwise.
  d) Use your graph to estimate a value of the effective diffusion coefficient D_{eff}.

(3.6) Droplets of molten silver, with radius $100\,\mu$m, are observed under a microscope as the temperature is lowered below the melting point. A large number of droplets all solidify 227 K below the melting point.

a) Assuming that these droplets are solidifying by homogeneous nucleation, calculate the solid/liquid interfacial energy of silver. You may assume that a droplet solidifies when it contains, on average, one nucleus of the critical size, and that during the experimental time-scale, each atom makes 5×10^{13} attempts to form a nucleus.

b) Recent computer simulations suggest that the interface between a crystal and its melt, rather than being atomically sharp, is between 5 and 10 atomic spacings broad. What are the implications of this result for classical nucleation theory? Illustrate your answer by calculating the classical critical nucleus size in part a.

[Data required: Silver: relative atomic mass = 108, density = $10.49\,\text{g cm}^{-3}$, melting point 1234 K, latent heat of fusion = $1.1 \times 10^9\,\text{J m}^{-3}$.]

Colloidal dispersions

4

4.1 Introduction

A **colloidal dispersion** is a heterogeneous system in which particles of solid or droplets of liquid with dimensions of order $10\,\mu\text{m}$ or less are dispersed in a liquid medium. Familiar examples of such systems include industrial and household products such as paints and inks, food products such as mayonnaise and ice cream, and biological fluids such as blood and milk.

Colloidal dispersions are characterised by an extremely high area of interface; for example, if 1 kg of polymer is dispersed in water in the form of spheres of 200 nm radius (this is roughly what a 5 litre tin of emulsion paint contains, in addition to pigment and some other ingredients), the total area of of interface between water and polymer is around $15\,000\,\text{m}^2$. Associated with this area of interface is a substantial amount of interfacial energy, and one has to ask the question why the polymer particles do not combine to form larger aggregates to reduce this interfacial energy.

Thus understanding the **stability** or otherwise of a colloidal dispersion is a central issue. Gravity is one force which may destabilise a dispersion; if the dispersed particles are less dense than the dispersing fluid, they will tend to rise to the surface, or **cream**, as indeed the fat droplets in untreated full-cream milk will tend to do. A denser dispersed phase, in contrast, will tend to sediment. Opposing this tendency is the Brownian motion of the particles; as the dispersed particles become smaller the size of the gravitational force decreases until it may be in effect overcome by the random thermal motion of the particles. Full-cream milk is **homogenised** by forcing it through a small nozzle to reduce the size of the fat globules.

If the particles are small enough to minimise the effect of gravity in destabilising the dispersion, we still need to consider what happens if particles are able to collide with each other in the course of their random, Brownian motion. When colloidal particles are able to come into contact they will stick together irreversibly; as time goes on larger and larger assemblies of particles will be formed in a process known as **aggregation**. If we are to render a colloidal dispersion stable against aggregation, we must modify the forces acting between the colloidal particles, which are normally attractive, to make the particles repel each other. This can be done by exploiting electrostatic forces, in **charge stabilisation**, or by modifying the interfaces by attaching polymer chains to them, in **steric stabilisation**.

The flow properties of colloidal dispersions are often unusual and distinctive. The first effect of adding solid particles to a liquid is to increase the effective viscosity of the dispersion, but pronounced non-Newtonian effects are also apparent, including both shear thinning (a very useful property in a paint, for example) and shear thickening. Indeed, some colloidal dispersions can exhibit solid-like behaviour, with a finite (albeit usually rather small) shear stress.

4.2 A single colloidal particle in a liquid—Stokes' law and Brownian motion

4.2.1 Stokes' law

If one takes a solid sphere and drops it in a fluid, it accelerates under gravity until the drag force balances the gravitational force and it attains terminal velocity. Fluid mechanics tells us (see, for example, Faber (1995)) that the mechanism causing the drag depends on the value of a dimensionless group called the **Reynolds number**; for a sphere of radius a moving with velocity v in a liquid of viscosity η and density ρ the Reynolds number Re is given by

$$Re = \frac{\rho v a}{\eta}. \tag{4.1}$$

The Reynolds number measures the relative importance of inertia and viscosity in providing the mechanism of drag; for smaller particles, lower velocities, and larger viscosities it is direct viscous effects that increasingly dominate. For particles on a colloidal scale, the Reynolds number is very low, and in this viscous-dominated regime the drag force F_S is given by Stokes' law:

$$F_S = 6\pi \eta a v. \tag{4.2}$$

For an isolated sphere whose density differs from the density of the liquid by $\Delta \rho$, the gravitational force F_g is given by

$$F_g = \frac{4}{3}\pi a^3 \Delta \rho g, \tag{4.3}$$

so the terminal velocity v_t, when the drag force is balanced by the gravitational force, is given by

$$v_t = \frac{2a^2 \Delta \rho g}{9\eta}. \tag{4.4}$$

4.2.2 Brownian motion and the Einstein equation

If one examines a dilute suspension of colloidal spheres in water or some other liquid, one soon sees that the static picture of classical fluid mechanics is incomplete. The experiment can be easily performed by looking at a dilute suspension of a polymer colloid, consisting of spherical particles a micrometre or less in size, in an optical microscope: what one sees is that each particle moves about with a continuous but random jiggling motion. This motion is known as **Brownian motion**, after Robert Brown, a botanist who discovered the phenomenon in 1827 while looking at plant pollen under the microscope.

Most readers will recall from elementary science how this phenomenon can be used to illustrate the existence of atoms and molecules in constant thermal motion. We can think of the colloidal particle as being constantly bombarded by the random impacts of the molecules of the liquid. Because these collisions are random, in the long run the total force acting on the particle as a result of the collisions is zero, but at any one time there will be more collisions on one side of the particle than another, and the result is that there is a constantly fluctuating net force. It is the motion of the particle in response to this fluctuating force that we see as Brownian motion.

How can this Brownian motion of a particle be characterised? It is easy to see that the motion has the character of a **random walk**. If we plot out the position of the particle at successive small intervals of time, we will see that at each stage its position will have moved, but that the directions of successive steps are not correlated. In such a random walk the mean of the total displacement is always zero, but the mean value of the square of the displacement is proportional to the number of steps, and thus the time. Thus, if the displacement vector after time t is $\mathbf{R}(t)$ then

$$\langle (\mathbf{R}(t))^2 \rangle = \alpha t. \tag{4.5}$$

We can find the value of the constant α (which is related to the diffusion constant D) using an argument due to Einstein and Smoluchowski. We can write the equation of motion of the particle in the following form:

$$m \frac{d^2 \mathbf{R}}{dt^2} + \xi \frac{d\mathbf{R}}{dt} = F_{\text{random}}. \tag{4.6}$$

Here we assume that there is a drag force on the particle proportional to the velocity, with a drag coefficient ξ. For a sphere of radius a in a liquid of viscosity η this would be given by Stokes' law:

$$\xi = 6\pi \eta a. \tag{4.7}$$

The applied force is the random force F_{random} resulting from the collisions of the water molecules with the sphere.

Clearly as the forces are random each space dimension must behave in the same way, so if we find an expression for the mean of x^2 the y and z directions must behave in the same way: $\langle x \rangle^2 = \langle y \rangle^2 = \langle z \rangle^2$ and $\langle \mathbf{R}^2 \rangle = 3\langle x^2 \rangle$. We can write $d(x^2)/dt = 2x(dx/dt)$, so multiplying eqn 4.6 by x and rearranging gives

$$\frac{\xi}{2} \frac{d(x^2)}{dt} = x F_{\text{random}} - mx \frac{d^2 x}{dt^2}. \tag{4.8}$$

We need to rewrite this using an identity for $x(d^2 x/dt^2)$:

$$x \frac{d^2 x}{dt^2} = \frac{d}{dt} \left(x \frac{dx}{dt} \right) - \left(\frac{dx}{dt} \right)^2. \tag{4.9}$$

Making this substitution, and taking the average of each term, we have

$$\frac{\xi}{2} \frac{d\langle (x^2) \rangle}{dt} = \langle x F_{\text{random}} \rangle - m \frac{d}{dt} \left\langle x \frac{dx}{dt} \right\rangle - m \left\langle \left(\frac{dx}{dt} \right)^2 \right\rangle. \tag{4.10}$$

Because the direction of the random force is uncorrelated with the position of the particle, the first term on the right hand side of the equation is zero. The second term is also zero; for a similar reason, there is no correlation between the object's position and its velocity. We can rewrite the last term using the theorem of equipartition of energy; for any object in thermal equilibrium at temperature T we can write $mv_x^2/2 = k_BT/2$. Thus we find

$$\frac{d\langle (x^2)\rangle}{dt} = 2\frac{k_BT}{\xi}, \tag{4.11}$$

giving us for the total mean squared displacement

$$\langle (\mathbf{R})^2\rangle = \frac{6k_BT}{\xi}t. \tag{4.12}$$

The motion of the particle is **diffusive**, with a diffusion coefficient D given by the **Einstein formula**,

$$D = \frac{k_BT}{\xi}. \tag{4.13}$$

For the case of a sphere diffusing in a liquid, we have from eqn 4.2 $\xi = 6\pi\eta a$, giving us for the diffusion coefficient D_{SE}

$$D_{SE} = \frac{k_BT}{6\pi\eta a}. \tag{4.14}$$

This relation is known as the **Stokes–Einstein** equation.

The Stokes–Einstein relation forms the basis of a very direct way of measuring Boltzmann's constant, by tracking the Brownian motion of colloidal particles of known diameter in a fluid of known viscosity. This was first done by Jean Perrin in the first decade of the twentieth century, in experiments which provided the final demonstration of the validity of the atomic hypothesis. Nowadays the relation is more often used to determine the size of unknown colloidal particles; diffusion coefficients can now be measured using **dynamic light scattering**, and the radii of the particles deduced from eqn 4.14.

4.3 Forces between colloidal particles

4.3.1 Interatomic forces and interparticle forces

A colloid is, by definition, a system with a large amount of surface. Associated with this surface is an energy whose origin lies in the fundamental forces between the atoms or molecules that make up the dispersed phase. To see the orders of magnitude involved, consider how much energy is involved in the surfaces of an everyday colloid such as an emulsion paint. Here, the particles are spheres of polymer with a radius of order 100 nm. The interfacial energy of a relatively polar polymer such as the acrylics used in paints with water might be of order 20 mJ m^{-2}, so the interfacial energy associated with each particle is about 2×10^{-15} J. This is many orders of magnitude bigger than the value of k_BT at room temperature, so on the face of it we would expect such a dispersion to be highly unstable. To prevent the colloid from quickly aggregating to form a single polymer mass there must be other forces between the particles that prevent them coalescing. All such forces in colloidal systems are fundamentally electrostatic in origin, but despite this common origin the manifestations of these forces can be so different that it is convenient to treat different types of force as separate entities. Even in the apparently straightforward situation, where one has two surfaces that are similarly electrostatically charged, the resulting forces are not primarily due to simple electrostatic repulsion. Instead, it is the behaviour of the medium separating the surfaces, and in particular of the free charges dissolved in the solvent, that determines the force between the surfaces.

4.3.2 Van der Waals forces

Perhaps the simplest force that we need to consider is the van der Waals force. We know that there is an attractive force between any pair of atoms or molecules, even when the atoms are uncharged and have no dipole moment. The origin of this force is ultimately quantum mechanical, arising from the interaction between fluctuating dipoles in each of the atoms. When two surfaces interact, there is a resultant force arising between the mutual interaction of all the pairs of molecules on the opposite surfaces. The estimation of the force between two uncharged atoms is a straightforward application of quantum mechanical perturbation theory, described in many textbooks. The result at this level of approximation is that the potential $V(r)$ varies as the inverse sixth power of the separation r. Specifically

$$V(r) = -\frac{3}{4}\left(\frac{1}{4\pi\epsilon_0}\right)^2\frac{\alpha^2}{r^6}\hbar\omega, \tag{4.15}$$

where α is the polarisability and the ionisation energy is $\hbar\omega$. To proceed from a knowledge of the forces between two atoms to the total force between two macroscopic bodies, the simplest approach simply sums up the interactions between all pairs of atoms in each of the bodies. Thus if we write the potential between two atoms as $-C/r^6$, we find the total potential $U(h)$ between two macroscopic bodies as a function of their separation h by a double integral over the volumes of each of the bodies. Thus

$$U(h) = \int\int -\frac{C}{r^6}\rho_1 dV_1 \rho_2 dV_2 \tag{4.16}$$

where the atomic number density of the volume element dV_1 of the first body is ρ_1.

For example, let us consider first a single atom a distance D away from a semi-infinite medium of density ρ (see Fig. 4.1). The interaction between the molecule and a ring of radius x whose centre is z away from the molecule is $-2\pi\rho x\, dx\, dz\, C/(z^2 + x^2)^3$. Thus the total interaction energy $w(D)$ is

$$w(D) = -2\pi\rho C \int_D^\infty dz \int_0^\infty \frac{x\, dx}{(z^2 + x^2)^3}$$

$$= -\frac{2\pi\rho C}{12D^3}. \tag{4.17}$$

If instead of a single atom we have a sheet of atoms of unit area and thickness dz at a distance z from the semi-infinite sheet, our energy per unit area

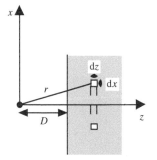

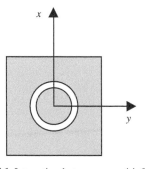

Fig. 4.1 Interaction between a semi-infinite medium and an atom at a distance D away from the surface. The total interaction energy is found by considering the interaction between the atom and an annulus of material and integrating.

is simply $-2\pi\rho^2 C/12z^3$. Thus we can find the total interaction per unit area between two semi-infinite sheets of the same material by another integration as

$$
\begin{aligned}
U(h) &= -\frac{2\pi\rho^2 C}{12} \int_h^\infty \frac{\mathrm{d}z}{z^3} \\
&= -\frac{A}{12\pi h^2},
\end{aligned}
\tag{4.18}
$$

where we have defined a constant $A = \pi\rho^2 C$, which is known as the **Hamaker** constant. The Hamaker constant is a material property with the dimensions of energy; for many materials its order of magnitude is around 10^{-19} J.

One important aspect of van der Waals forces between bodies is their magnitude. For two plates at contact (say at a separation of 0.2 nm) one finds an attractive energy per unit area of around 0.1 J m^{-2}, and a force per unit area of about 10^9 Pa, or 10 000 atmospheres. Even at a separation of 10 nm the force per unit area is still significant. It is the additive character of van der Waals forces that means that even though between two individual atoms the interaction is rather weak, between larger bodies the force can be very strong. If two macroscopic bodies can be brought together close enough to be approaching atomic scale contact over significant areas they will stick together rather strongly. In practice we do not observe this for most macroscopic solids because their surfaces are too rough to obtain enough intimate contact. However, close contact may be achieved for soft, deformable solids; this is largely why soft plastics like cling-film adhere so effectively to a wide variety of other solids.

This simple approach to calculating van der Waals forces between bodies suffers from two drawbacks:

1. The assumption of pairwise additivity of forces is strictly incorrect, because the magnitude and direction of the fluctuating dipoles in any pair of atoms is influenced not just by each other but by all the other atoms in the system.
2. For larger separations (say, of order 10 nm and greater), the effects of the finite speed of the propagation of fields arising from the fluctuating dipoles become significant. For large values of the separation r, only dipole fluctuations with a frequency low compared to c/r, where c is the speed of light, result in an induced force. As a result of this so-called **retardation** effect the potential at large distances varies as r^{-7} rather than r^{-6}. Correspondingly, after pairwise integration we expect the potential between two semi-infinite sheets for large values of separation h to vary as h^{-3} rather than h^{-2}.

A more powerful approach to calculating forces between bodies relies on applying quantum field theory to determine the properties of the electromagnetic field between the two bodies. The full application of this approach—the **Lifshitz theory**—is beyond the scope of this book, but we can get a flavour of the physics involved by considering a very simple situation: the force between two perfect conductors in a vacuum at a temperature of 0 K. This force is sometimes known as the **Casimir effect**.

The origin of the Casimir effect lies in the zero-point energy of the electromagnetic field in empty space. Recall, for example, the derivation of the energy spectrum of black body radiation—we consider each possible standing wave within a cavity as an independent oscillator obeying Bose–Einstein statistics. Even at absolute zero, each possible standing wave mode has a zero-point energy of $\hbar\omega/2$. There are an infinite number of possible standing wave modes in any region of space, so the total zero-point energy—the **vacuum energy**—is itself infinite. If one has two perfectly conducting metallic plates, the range of allowed standing wave modes in the space between them is modified, because boundary conditions at the surface of the plates need to be satisfied. Remarkably, although the total zero-point energy of the electromagnetic field is infinite both with and without the plates, the difference in the total energies of the two situations is finite, and can be easily calculated.

Consider two perfectly conducting, parallel, plates of unit area, separated by a distance l. At absolute zero we can write the total energy $E(l)$ of the electromagnetic field between the plates in terms of the sum of all the zero-point energies:

$$E(l) = \frac{\hbar}{2} \sum_{k_1,k_2,n} \omega(k_1, k_2, n), \qquad (4.19)$$

where the frequencies of the modes are given by

$$\omega(k_1, k_2, n) = c \left(k_1^2 + k_2^2 + \frac{n^2\pi^2}{l^2} \right)^{1/2}. \qquad (4.20)$$

Here k_1 and k_2 are the components of wavevectors in the x and y directions, taken to be in the plane of the plates, and c is the velocity of light. We can find the total number of possible values of k_1 and k_2 by integrating over two-dimensional k-space to find

$$E(l) = \frac{\hbar c}{2} \frac{2}{2\pi^2} \int_{-\infty}^{\infty} \int_{-\infty}^{\infty} \left[\sum_{n=1}^{\infty} \left(k_1^2 + k_2^2 + \frac{n^2\pi^2}{l^2} \right)^{1/2} \right.$$
$$\left. + \frac{1}{2}(k_1^2 + k_2^2)^{1/2} \right] dk_1 dk_2. \qquad (4.21)$$

Here we have taken into account that there are two modes corresponding to two polarisations for each integral value of n, and also a single mode for $n = 0$. We can slightly simplify this by integrating over the possible values of the magnitude of the wavevector in the plane of the plates, $\kappa = (k_1^2 + k_2^2)^{1/2}$, to give

$$E(l) = \frac{\hbar c}{2\pi} \int_0^{\infty} \sum_{n=0}^{\infty}{}' \left(\kappa^2 + \frac{n^2\pi^2}{l^2} \right)^{1/2} \kappa \, d\kappa, \qquad (4.22)$$

where the prime on the summation reminds us to multiply the $n = 0$ term by $1/2$.

How is the energy different in the absence of the plates? Instead of definite quantised values of n arising from the boundary conditions at the plates (see Fig. 4.2), the perpendicular component of the wavevector can now take a continuum of values. The effect of this on the calculation of the total zero-point energy is to replace the sum of eqn 4.22 by an integral, so we can write

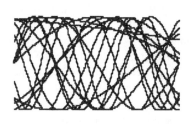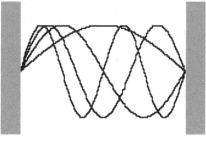

Fig. 4.2 Sketch in one dimension of some of the possible standing wave modes in free space (left), and in the space between perfectly conducting metal plates (right). The difference in zero-point energies between the two situations gives rise to the attractive Casimir force between the plates.

the zero-point energy contained within the same volume in the absence of plates, $E_{\text{free}}(l)$, as

$$E_{\text{free}}(l) = \frac{\hbar c}{2\pi} \int_0^\infty \kappa \, d\kappa \int_0^\infty \left(\kappa^2 + \frac{n^2\pi^2}{l^2} \right)^{1/2} dn. \tag{4.23}$$

Clearly both eqns 4.22 and 4.23 predict infinite energies, but it turns out to be possible to show that the difference between the two expressions remains finite, and it is this difference that provides the attractive Casimir potential between the two plates, $U_C(l) = E(l) - E_{\text{free}}(l)$. To do this we introduce a new function, $S(\delta, n)$, given by

$$S(\delta, n) = \int_0^\infty \kappa \, d\kappa \left(\kappa^2 + \frac{n^2\pi^2}{l^2} \right) \exp\left[-\left(\kappa^2 + \frac{n^2\pi^2}{l^2} \right) \delta^2 \right]. \tag{4.24}$$

In terms of this function we can write expressions for our energies in the presence and absence of the plates:

$$E(l) = \lim_{\delta \to 0} \frac{\hbar c}{2\pi} \sum_{n=0}^\infty {}' S(\delta, n), \tag{4.25}$$

and

$$E_{\text{free}}(l) = \lim_{\delta \to 0} \frac{\hbar c}{2\pi} \int_{n=0}^\infty S(\delta, n) \, dn. \tag{4.26}$$

To find the change in energy that results from the presence of the plates, we simply need to evaluate the difference between a series and a corresponding integral. The Euler–Maclaurin formula gives us a way of doing just this. Using this, we can write

$$\sum_{n=0}^\infty {}' S(\delta, n) = \int_{n=0}^\infty S(\delta, n) dn + \frac{1}{12} \frac{dS}{dn} \Big|_{n=0}^{n=\infty}$$
$$+ \frac{1}{720} \frac{d^3 S}{dn^3} \Big|_{n=0}^{n=\infty} + \cdots. \tag{4.27}$$

The derivative of $S(\delta, n)$ with respect to n is found to be

$$\frac{dS(\delta, n)}{dn} = n^2 \left(\frac{\pi}{l} \right)^3 \exp(-n^2\pi^2\delta^2/l^2), \tag{4.28}$$

so if we evaluate the third derivative, put in the limits for n, and set $\delta = 0$, we finally arrive at the desired answer:

$$U_C(l) = E(l) - E_{\text{free}}(l) = -\frac{\hbar c \pi^2}{720 l^3}. \tag{4.29}$$

We note that the Casimir force that is derived from this potential is always attractive, and that the potential varies as the inverse third power of the separation. This corresponds to the variation that is predicted for van der Waals forces in the retarded regime. Indeed, the Casimir force is identical to the van der Waals force in this limit of perfect conductors and a temperature of absolute zero. In more realistic situations, a number of modifications must be made.

1. Real metals are not perfect conductors, and the electric field is not, therefore, zero exactly at their surfaces. Instead the field penetrates a certain **skin depth**, of order 100 nm. The Casimir analysis only applies for distances large compared to the skin depth; for smaller distances we expect to recover a potential varying as the inverse square of the separation, as predicted for non-retarded van der Waals forces.
2. More generally, the response of a material to an applied electric field is given by its frequency-dependent complex dielectric susceptibility, and it is this quantity that must be used to determine the way in which matter modifies the permitted electromagnetic modes.
3. At finite temperatures, for each mode we have to consider not just the zero-point energy, but also the thermal energy of each mode as given by Bose–Einstein statistics. This results in a contribution to the potential proportional to $k_B T / l^2$. However, the contribution to the total potential of this entropic force is usually rather minor.[1]

Full calculations of the van der Waals force between two bodies are complicated and require detailed knowledge of the electrical properties of the materials. The most usual approach is to use the expression of eqn 4.18,

$$U(h) = -\frac{A}{12 \pi h^2}, \tag{4.30}$$

using various approximate expressions derived from the Lifshitz theory for the Hamaker constant A.

Van der Waals forces, and their special case, the Casimir force, are the direct manifestation of the fact that the electromagnetic field between two bodies has an associated zero-point energy whose magnitude varies with the separation of the bodies. The force can be thought of as a direct manifestation of the vacuum fluctuations in the space between the bodies. This is a remarkably different conception to the original notion of van der Waals force as arising between the fluctuations of two dipoles, but nonetheless these two very different-looking approaches are describing the same phenomenon. This identity of the Casimir force and the van der Waals force is not always recognised, perhaps because of a reluctance to accept that such a subtle effect, relying on the mysteries of the quantum field, underlies so many very mundane phenomena. Nonetheless, it is a fact that every time one wraps one's sandwiches in cling-film one is relying on the subtleties of vacuum fluctuations and virtual photons to stop the crumbs falling out.

[1] An important exception to this comment occurs when hydrocarbons interact across water, when the entropic term may be comparable to or even larger than the term arising from zero-point energy.

4.3.3 Electrostatic double-layer forces

Many surfaces are charged, and so one may expect direct electrostatic interactions to be important in determining the forces between colloidal objects with charged surfaces. However, when the objects are suspended in water, dissolved ions are always present, and the interaction of the charged bodies with the free ions profoundly modifies the nature of the electrostatic interaction. In particular, the electrostatic interactions are **screened** by dissolved ions; rather than a direct Coulomb interaction between two charged bodies, one finds a screened Coulomb interaction which exponentially decays in strength with distance.

Suppose one has a surface that is ionised—this may be because some chemical groups at the surface ionise, or some ions from solution adsorb. Overall charge neutrality will be maintained by a layer of **counterions** which will be attracted to the surface by the electrostatic field. Some of these counterions may be tightly bound to the surface (this layer of tightly bound ions is known as the **Stern** layer), but more will form a diffuse concentration profile away from the surface.

How can we determine the form of this concentration profile? There will be an electrostatic potential $\psi(x)$ at a distance x from the surface, and the density of ions $n(z)$ will be determined by the Boltzmann equation

$$n(z) = n_0 \exp\left(\frac{-ze\psi(x)}{k_B T}\right), \tag{4.31}$$

where the charge of the ions is ze.

Now the potential $\psi(x)$ is itself determined by the distribution of net charge $\rho(z)$ by the Poisson equation

$$\rho(z) = -\epsilon\epsilon_0 \left(\frac{d^2\psi}{dx^2}\right). \tag{4.32}$$

In the simplest case where the only ions present are the counterions needed to balance the charge of the surface, $\rho = ze$, and we can combine these two equations to give the **Poisson–Boltzmann** equation:

$$\frac{d^2\psi}{dx^2} = -\left(\frac{zen_0}{\epsilon\epsilon_0}\right)\exp\left(-\frac{ze\psi}{k_B T}\right). \tag{4.33}$$

This is an important equation that is also met in plasma physics and the study of electrons in solids.

In one very common case the surface is in contact with a solution of an electrolyte which is a solution of a univalent salt such as sodium chloride. Now we have the concentrations of both negative and positive ions to consider; taking these concentrations to be n_+ and n_- we have

$$n_\pm = n_0 \exp\left(\mp\frac{ze\psi}{k_B T}\right), \tag{4.34}$$

where n_0 is the ionic concentration in bulk solution. The net charge density is given by

$$\rho = ze(n_+ + n_-) \tag{4.35}$$

so

$$\frac{d^2\psi}{dx^2} = \frac{2zen_0}{\epsilon\epsilon_0} \sinh\left(\frac{ze\psi}{k_B T}\right).$$ (4.36)

We need to solve this equation subject to boundary conditions; for an isolated plate in solution these are that both the potential ψ and its gradient $d\psi/dx$ approach zero as x approaches infinity.

If the potential is small we can make the approximation $\sinh(ze\psi/k_B T) \approx (ze\psi/k_B T)$; in this limit (known as the Debye–Hückel approximation) eqn 4.36 has the solution

$$\psi(x) = \psi_0 \exp(-\kappa x),$$ (4.37)

where κ has the value

$$\kappa = \left(\frac{2e^2 n_0 z^2}{\epsilon\epsilon_0 k_B T}\right)^{1/2}.$$ (4.38)

Thus in an electrolyte, electric fields are **screened**. The length which characterises this screening, κ^{-1}, is known as the **Debye screening length**. At distances much greater than the Debye screening length, which is inversely proportional to the square root of the concentration of salt in the electrolyte, the strength of the direct electrostatic interaction between charged objects rapidly falls to zero.

The variation of ionic concentrations is shown schematically in Fig. 4.3. For monovalent salts the Debye screening length in water is given by $\kappa = 0.304\,I^{-1/2}$ nm, where I is the concentration of salt in moles/l. Thus even for relatively modest salt concentrations electrostatic effects are strongly screened.

What happens when two equally charged plates are brought together? As one might expect, there is a repulsive force between them, but less obviously the origin of this force is not due to the direct effect of electrostatics. The combination of the charged surface and the attracted counterions must overall be charge neutral, so this cannot lead to any repulsive force; instead it is the

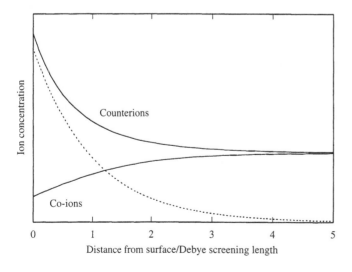

Fig. 4.3 Distribution of ions near a charged surface, according to Debye–Hückel theory. The dotted line illustrates the form of the potential near the surface.

excess osmotic pressure of the counterions in the gap between the plates that lead to a repulsive force. It can be shown that in the limit of large separations and low surface potentials the repulsive force per unit area P between plates at separation D is given by

$$P = 64 k_B T n_0 \tanh^2 \left(\frac{ze\psi_0}{4k_B T} \right) \exp(-\kappa D). \qquad (4.39)$$

This equation emphasises that forces of electrostatic origin in electrolyte solutions are short ranged, with the Debye screening length κ determining that range. The dependence of κ on the concentration of ions means in practice that colloidal stability is profoundly affected by the salt concentration.

4.3.4   Stabilising polymers with grafted polymer layers

One of the most important practical methods for stabilising colloids is by coating the particle with a polymer layer. The principle of this approach is illustrated in Fig. 4.4; chains are attached by one end to the particle surface and stick out into the solution. If two particles approach one another, the concentration of polymer solution inside the gap increases. This leads to an increase in osmotic pressure, causing a repulsive force.

We will return to the details of this process when we have considered the thermodynamics of polymers in solution in the next chapter. For the moment, we make the following qualitative remarks.

- For the polymer to be effective at stabilising the colloid, the solvent must be a good solvent for the polymer. If there is an effective attractive interaction between polymer segments that would in a bulk solution of the polymer lead to phase separation, then the polymer will lead to an attractive interaction between the particles.
- The range of the interaction is governed by the distance from the surface that the polymer chains extend. As we will see in Section 5.3.6, this is controlled by the length of the polymer chain, the density at which chains are grafted, and the strength of the interaction between polymer segments and the solvent.

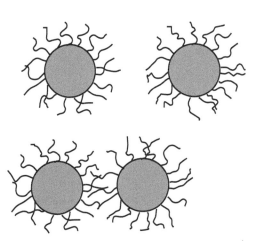

Fig. 4.4 Stabilisation of colloids with grafted polymers. When the particles come close enough for the grafted polymers to overlap, a local increase in polymer concentration leads to a repulsive force of osmotic origin.

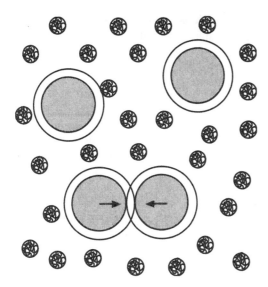

Fig. 4.5 The depletion interaction. Polymer coils are excluded from a depletion zone near the surface of the colloidal particles; when the depletion zones of two particles overlap there is a net attractive force between the particles arising from unbalanced osmotic pressures.

- The chains can be attached to the surface either by chemical bonds or by physical interactions. In either case, the strength of the bond anchoring the polymer to the surface needs to be rather greater than $k_B T$, otherwise there will be a tendency for the chain to become detached from the surface.

4.3.5 Depletion interactions

The final interaction we consider is, like the electrostatic interactions and polymer stabilisation, ultimately due to osmotic pressure, but in this case the interaction is **attractive** rather than repulsive. **Depletion interactions** arise whenever the solution contains, in addition to the suspended particles, other particles intermediate in size between the suspended particles and the size of the solvent molecules. The most common case occurs when the suspension contains a dissolved polymer which does not adsorb onto the surface of the particles.

The situation is illustrated in Fig. 4.5. The polymer molecules, depicted here as spheres, are excluded from a region of thickness L away from the surface of the particles—the depletion zone. As the particles approach, the depletion zones overlap, with the result that there is a volume of solution between the particles in which the concentration of polymer molecules is less than it is in the bulk solution. This means that the difference in osmotic pressure between the bulk solution and the depletion zone leads to a force pushing the particles together.

For a dilute solution of polymers, or indeed any other particles, which do not interact, the osmotic pressure P_{osm} is given simply by the ideal gas expression

$$P_{osm} = \frac{N}{V} k_B T, \qquad (4.40)$$

where there are N polymer molecules in volume V of solution. The net interaction potential between the particles F_{dep} is simply

$$F_{dep} = -P_{osm} V_{dep}, \qquad (4.41)$$

where V_{dep} is the total volume between the particles from which the polymers are excluded. For two spheres of radius a, at a centre-to-centre separation r, simple geometry gives

$$V_{\text{dep}} = \frac{4\pi}{3}(a + L^3)\left(1 - \frac{3r}{4(a + L)} + \frac{r^3}{16(a + L)^3}\right). \qquad (4.42)$$

The depletion interaction is never very large, but it is always attractive, and the depth of the well can become comparable to $k_{\text{B}}T$; thus increasing the strength of the depletion interaction by addition of free polymer can lead to phase separation or aggregation in colloidal systems.

4.4 Stability and phase behaviour of colloids

We have seen how a variety of interactions between colloidal particles may lead to the particles either repelling or attracting each other; these interactions lead to interesting phase behaviour. In some ways this phase behaviour can be considered to be analogous to the phase behaviour of matter, with the colloidal particles taking the role of atoms or molecules. One can envisage a phase transition from a solid-like phase to a liquid-like phase, driven by attraction between the particles. More usually, however, the attractive energy is much greater than $k_{\text{B}}T$. Rather than having an equilibrium between a dense liquid-like phase and a gas-like phase, whenever two particles meet, they stick irreversibly. This leads to the formation of open flocs which eventually fall out of suspension (flocculate).

We can summarise some of the ways in which the interaction between colloidal particles can be changed from repulsive to attractive:

- we can **add salt** to an electrostatically stabilised colloid, reducing the Debye screening length and decreasing the magnitude of the electrostatic repulsion relative to the van der Waals attraction;
- we can **add poor solvent** to a polymerically stabilised colloid; the resulting attractive polymer/polymer interactions will lead to a net attraction between the colloid particles;
- we can **physically or chemically remove** grafted polymer chains from the surface of the colloidal particles;
- we can **add non-adsorbing polymer** to cause an increase in the size of the depletion interaction.

4.4.1 Crystallisation of hard-sphere colloids

When the forces between colloidal particles are repulsive at all separations one has a stable suspension. If the particles are spherical with a rather narrow size distribution, then as one increases the concentration of particles (e.g. if the dispersing liquid is allowed to evaporate) one finds a remarkable transition from a disordered arrangement of particles, analogous to a liquid, to a crystalline packing of the particles. These **colloidal crystals** have true long-range order; often the diameter of the colloidal particles falls between $100\,\text{nm}$ and $1\,\mu\text{m}$, and in this case the crystal will diffract light, resulting in striking opalescent

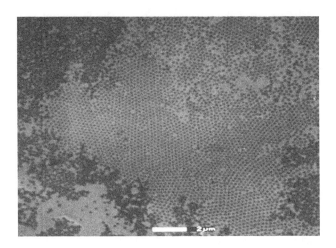

Fig. 4.6 A water-borne latex suspension imaged by environmental scanning electron microscopy, showing the formation of ordered regions. Reprinted with permission from He and Donald (1996). Copyright 1996 American Chemical Society.

interference effects. In fact, the gemstone opal is a natural colloidal crystal composed of submicrometre-sized silicon dioxide particles. The systems typically studied in the laboratory are composed of submicrometre polymer particles produced by a so-called emulsion polymerisation process, which yields rather a narrow particle size distribution. The particles are stabilised against coagulation by coating them with a layer of short polymer chains grafted to the surface. In fact, these systems are very similar to the latices produced on a vast industrial scale to form the basis of emulsion paints and water-based varnishes. An environmental scanning electron micrograph of one such material is shown in Fig. 4.6, showing clearly the tendency of the particles to form close-packed ordered regions.

A good starting point for understanding the phenomenon of colloidal crystallisation is to be found in a very simple model developed by theorists to understand liquids—the **hard-sphere** model. As its name suggests, in this model we consider an assembly of perfect spheres which interact via a potential which is zero except where two spheres overlap, in which case it is infinite. This model is in fact rather a good approximation to a sterically stabilised colloid in which the thickness of the stabilising polymer layer is much less than the radius of the particles.

We know that the maximum density one can achieve in packing hard spheres is obtained when they are arranged in a regular close-packed structure, in which case the volume fraction of spheres is 0.7404. If the spheres are packed randomly as closely as possible, then the maximum volume fraction obtainable (**random close packing**) can be shown to be around 0.63. So one can see that at very high volume fractions simple packing constraints force the spheres to take up a close-packed crystalline structure.

However, this provides only a very partial explanation of colloidal crystallisation. What is surprising is that crystals appear at a much lower volume fraction of spheres than that for either regular close packing or random close packing. In fact, at a volume fraction of 0.494 there is an abrupt transition to a crystal with a volume fraction of 0.545. This is a true phase transition; if one prepares a suspension with an intermediate volume fraction it will separate into two coexisting phases.

What is the origin of this phase transition? The first point to note is that temperature does not directly enter into the problem. Because the only possible energies for any configurations of the sphere are either zero (if no spheres overlap) or infinity (if two spheres do overlap) the Boltzmann weights for any configuration can only be either unity or zero, with no dependence on temperature. So the volume fractions of the coexisting liquid and solid phases are completely independent of temperature. The transition between solid and liquid must be driven entirely by **entropy**.

This seems paradoxical. How can it be that for some volume fractions, a regular crystal with long-range order can have a higher entropy—that is, be more disordered—than a random, liquid-like, arrangement? The reason is related to the difference between random and regular close-packing densities. In the crystal state, we lose entropy compared to the amorphous state by virtue of the long-range order, but because crystalline packing is more efficient than random packing each individual sphere has more space locally to explore, and thus has a higher entropy in the crystalline state compared to the amorphous state.

Another way of thinking about this involves the idea of **excluded volume**. The fact that two spheres cannot overlap leads to an effective repulsive force between spheres of entropic origin. We can understand this by remembering how excluded volume is dealt with in the van der Waals theory of non-ideal gases. Recall that for a perfect gas one can write the entropy per atom S_{ideal} of N atoms in a ideal V as

$$S_{ideal} = k_B \ln \left(a \frac{V}{N} \right) \tag{4.43}$$

where a is a constant. Now, if the gas atoms have a finite volume b this reduces the volume accessible to any given atom from V to $V - Nb$, and we must modify our expression for the entropy accordingly:

$$S = k_B \ln \left(a \frac{(V - Nb)}{N} \right). \tag{4.44}$$

We can rewrite this in terms of S_{ideal} as

$$S = S_{ideal} + k_B \ln \left(1 - \frac{bN}{V} \right), \tag{4.45}$$

and if the volume fraction of atoms is low we can expand the logarithm to yield

$$S = S_{ideal} - k_B \left(\frac{N}{V} \right) b, \tag{4.46}$$

with a corresponding free energy given by

$$F = F_{ideal} + k_B T \left(\frac{N}{V} \right) b. \tag{4.47}$$

Thus there is an effective repulsion between the atoms, and in the case of a hard-sphere colloid it is an effective interaction of this kind that causes the particles to arrange themselves on a crystal lattice.

Of course, at the volume fractions at which hard spheres crystallise the approximations used in this simple expression have long since broken down,

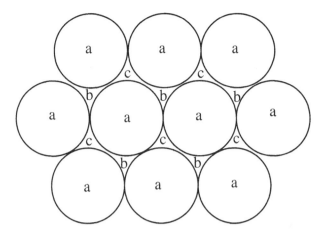

Fig. 4.7 A single close-packed layer, illustrating that there are two sites on which a second close-packed layer can be placed: b and c.

and the theory needs to be considerably refined to take into account the possibility that more than two particles may be interacting simultaneously, and to lift the mathematical approximations appropriate to a low particle volume fraction. Nonetheless developments of this approach form the basis of modern statistical mechanical theories of the hard-sphere fluid.

Before we leave the subject of crystallisation in hard-sphere systems, there is one subtle point that needs to be addressed. There are two different crystal structures which are close packed, namely face-centred cubic (FCC) and hexagonal close packed (HCP), and both have an identical maximum packing fraction. The difference between the two types of packing is best understood in terms of Fig. 4.7. Once one close-packed layer is laid down, there are two different ways in which subsequent layers can be placed. In the HCP structure, the sequence of layers alternates as ababab..., while in an FCC structure the sequence is abcabc....

Experimentally it is found that colloidal crystals as normally prepared have a random sequence of close-packed planes, corresponding to an HCP structure with a very large number of stacking faults. This is of some practical significance, as potentially colloidal crystals could have important applications in photonics, as materials with a so-called **photonic bandgap**. In such a material diffraction effects can lead to a situation in which light of a certain wavelength is unable to propagate in any direction. The existence of such bandgaps does, however, depend on the crystal structure being largely defect free. Quite recently, it has been found that there is a tiny difference in entropy between the FCC and HCP structure, and that it is the FCC structure that is always at equilibrium (Woodcock 1997, Mau and Huse 1999). Thus with careful preparation techniques it should be possible to prepare defect-free colloidal crystals with interesting optical properties.

4.4.2 Colloids with longer ranged repulsion

We have seen that we can produce a repulsive interaction between colloidal particles with a relatively long range either by having an electrostatic repulsion between the particles in a solution with a relatively low salt content, and thus a large Debye length, or by having long polymers grafted to the interface. Often

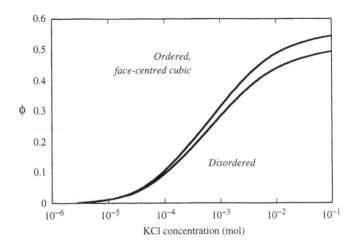

Fig. 4.8 Phase diagram for charged spheres in a polyelectrolyte solution as a function of the volume fraction of spheres ϕ and the concentration of salt, as calculated for spheres of radius 0.1 μm with surface charge $5000e$. After Russel *et al.* (1989).

the range of the repulsion is large compared to molecular sizes, but still small compared to the radius of the particle, and in this case the repulsive force can be treated as a perturbation of the hard-sphere interaction. Thus we anticipate a transition from a liquid state to a solid state, but this transition will occur at lower volume fractions than for a pure hard-sphere interaction. The effective hard-sphere radius of the colloid particles is greater than their physical radius because of the additional long-ranged repulsion.

This is illustrated in Fig. 4.8, which shows the predictions of a perturbation theory for the phase behaviour of the order–disorder transition for charged spheres in a solution of an electrolyte. At high concentrations of salt, the Debye screening length is very small and the order–disorder transition occurs at concentrations very close to those expected for ideal hard spheres. As the salt concentration is reduced, the Debye length—and thus the range of the repulsive interaction—increases and the volume fraction at which the order–disorder transition takes place is correspondingly reduced.

4.4.3   Colloids with weakly attractive interactions

If the interaction between colloidal particles is attractive, then we expect the system to undergo a transition to a **disordered, condensed phase**; this is analogous to a gas–liquid transition in a molecular system. The best way to realise this situation in practice is to add non-adsorbing polymer to a colloidal system that is stabilised by electrostatic interactions or by a grafted polymer layer; this creates a relatively weak attractive part of the potential whose magnitude depends on the volume fraction of added polymer. An example of the kind of phase diagram obtained by plotting the volume fraction of particles on one axis and the amount of added polymer—and thus the size of the attractive interaction—is sketched in Fig. 4.9. This diagram has been compared to experiment and found to be in agreement. One should compare this diagram to a phase diagram for a simple fluid plotted in the density/temperature plane, as sketched in Fig. 2.3. In fact the statistical mechanical theories which can be used to predict phase diagrams for simple fluids are easily adapted to predict the phase diagrams for colloidal dispersions.

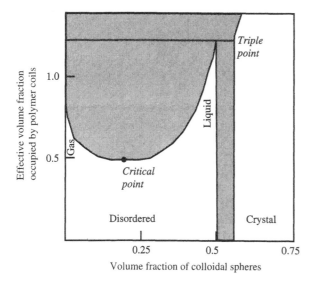

Fig. 4.9 Calculated phase diagram for a colloid of hard spheres with non-adsorbing polymer added to the solution. The ratio of the sizes of the colloidal spheres to the radii of the polymer molecules is 0.57. After Illett *et al.* (1995).

The qualitative features of this phase diagram are relatively straightforward to understand. When the attractive interactions are weak, we get a liquid–solid transition driven by the effective repulsion between the particles. This repulsion includes both the repulsion of entropic origin that underlies the excluded volume effect and any physical repulsion due to electrostatic or polymer-mediated interactions. The liquid–solid interaction is always first order; because there is a change in symmetry between the two phases the transition cannot take place gradually and thus there can be no critical point. When the attractive part of the potential becomes more important, we can have a gas–liquid phase separation which is analogous to the liquid–liquid unmixing transition discussed in Chapter 2. This transition can take place continuously, and thus there is a critical point. However, when the coexisting volume fraction for the liquid phase is large enough, this phase can lower its free energy even further by going over to an ordered state. Thus, according to the amount of added polymer, one can have a phase transition as a function of particle concentration from a gas to a liquid state, from a gas to a solid state, and indeed at one special condition there is a **triple point**, where gas, liquid, and solid coexist.

4.4.4 Colloids with strongly attractive interactions

The phase diagrams discussed in the section above are **equilibrium** phase diagrams; if one drives a system from the gas to the solid state by increasing the particle volume fraction, say, and if one subsequently adds more solvent then the transition should be completely reversible, and one should recover a low-density disordered state. If the depth of the attractive well in the potential curve is only a few times $k_B T$, equilibrium is easily reached. However, as the attraction becomes larger, it is more and more difficult to reach equilibrium: if two particles come into contact and stick then they take longer and longer to unstick and try another configuration, even if the first arrangement of particles does not correspond to the equilibrium state. This has a profound effect on the structure of the aggregates, as illustrated in Fig. 4.10.

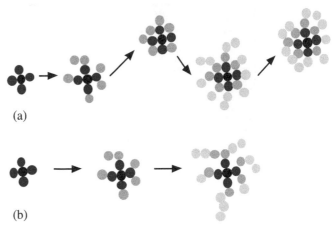

Fig. 4.10 Aggregation with and without rearrangement. In (a) the attraction is weak enough to allow the particles to rearrange following aggregation—this produces relatively compact aggregates. In (b) the attractive energy is so strong that once particles make contact, they remain stuck in this position. Particles arriving later tend to stick on the outside of the cluster, as access to its interior is blocked, resulting in much more open aggregates with a fractal structure.

If particles, once they have joined the aggregate, are able to rearrange, the resulting aggregate is likely to be rather compact. However, if once stuck the energy of attraction is too large for the particle to move again, the aggregate will be much more open in structure, because particles arriving at the aggregate later will tend to find access to the interior of the aggregate blocked by the earlier arrivals. In fact, the resulting structure is **fractal** in character. For a three-dimensional compact object the relationship between the size of the aggregate R and its mass M can be written $R \sim M^{1/3}$; for a fractal the corresponding relation is $R \sim M^{1/d_f}$, where d_f is a fractal dimension which is less than three but greater than one.[2]

In an idealised model of aggregation, known as **diffusion-limited aggregation**, particles are considered to diffuse randomly until they touch a member of the cluster, after which they stick without further movement. Computer simulations reveal that for this model $d_f \approx 1.71$, corresponding to very open, ramified structures. A more realistic model of aggregation would allow clusters themselves to aggregate; such **diffusion-limited cluster–cluster aggregation** models yield a slightly larger value of $d_f \approx 1.78$.

Experiments reveal that if aggregation takes place in circumstances in which aggregation is very fast, with very deep potential wells of attraction, open aggregates are found with fractal dimensions of around 1.75, while in experiments carried out in slower conditions, when some rearrangement is presumably permitted, fractal dimensions are somewhat larger.

4.5 Flow in concentrated dispersions

Adding particles to a liquid might be expected to produce a dispersion with a viscosity greater than that of the pure liquid; this is indeed what happens, but in addition to an increased viscosity concentrated dispersions can show pronounced non-Newtonian effects in their flow, in particular **shear thinning**.

[2] Of course, this relationship is valid for values of R between an upper limit, set by the overall size of the aggregate, and a lower limit defined by the size of the constituent particles. This is the difference between a mathematical fractal, for which such a relationship holds for all values of R, and a physical fractal, for which the relationship holds only for a limited range of lengths.

For a very dilute suspension of hard, spherical particles, with no long-ranged interactions, it can be shown that the viscosity is Newtonian, with a viscosity η related to the viscosity of the dispersing liquid η_0 by

$$\eta = \eta_0(1 + 2.5\phi), \tag{4.48}$$

where ϕ is the volume fraction of particles. But this works only for very dilute suspensions; for more concentrated suspensions we find both that the increase in viscosity with volume fraction is larger than predicted by eqn 4.48, and that the effective viscosity depends on **shear rate**.

The dependence on shear rate can be understood in terms of the effect of the flow on the **structure** of the dispersion. If the shear rate is large, then the structure of the dispersion will be perturbed by the application of the flow. The particles will rearrange themselves in order to minimise the amount of energy dissipated in the flow process, and this will reduce the effective viscosity at these high shear rates. On the other hand, if the shear rate is low, then the Brownian motion will restore the arrangement of the particles to their normal rest state.

We can use dimensionless analysis to estimate the shear rate at which we would expect the flow to perturb the structure of the dispersion. The diffusion coefficient D_{SE} for a single sphere of radius a is given by the Stokes–Einstein relation, eqn 4.14. A characteristic time for diffusion τ_D is given by the time taken for a sphere to diffuse a distance equal to its own size; thus

$$\begin{aligned}
\tau_D &= a^2/D_{SE} \\
&= \frac{6\pi\eta_0 a^3}{k_B T}.
\end{aligned} \tag{4.49}$$

The shear rate γ has dimensions of an inverse time, so a characteristic time corresponding to this shear rate τ_{shear} is given by

$$\tau_{shear} = \gamma^{-1}. \tag{4.50}$$

The ratio of these two characteristic times, τ_D/τ_{shear}, is a dimensionless number called the **Peclet number**, Pe, which expresses the relative importance of shear and Brownian motion. The Peclet number is given by

$$Pe = \frac{6\pi\eta_0 a^3 \gamma}{k_B T}. \tag{4.51}$$

At Peclet number much less than unity, the shear rate is low enough that Brownian motion is able to maintain the dispersion in its unperturbed state; at Peclet number greater than unity the shearing of the dispersion modifies its structure and Brownian motion is unable to restore this structure to the rest configuration on the timescale set by the shear rate.

On dimensional grounds, then, we would expect plots of relative viscosity as a function of shear rate for spherical particles at fixed volume fraction for a variety of dispersing liquids and particle sizes to fall onto a single curve when plotted against the Peclet number. This is indeed what is observed, and the resulting plot has a very characteristic shape, illustrated in Fig. 4.11. This shows a transition from a limiting viscosity at low shear rates to a lower limiting value of viscosity at high shear rates. The transition occurs over a broad range

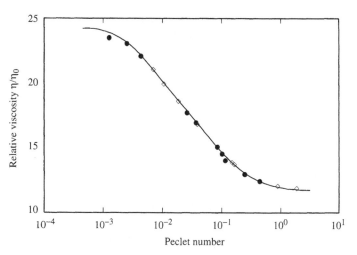

Fig. 4.11 Relative viscosity as a function of shear rate for model hard-sphere latices. The shear rate γ is plotted as the dimensionless combination, the Peclet number $Pe = 6\pi \eta_0 a^3 \gamma / k_B T$. The solid line is for polystyrene latices of radii between 54 and 90 nm in water; the circles are 38 nm polystyrene latices in benzyl alcohol, and the diamonds 55 nm polystyrene spheres in meta-cresol. Data from Krieger (1972), after Russel *et al.* (1989).

of shear rates—two or three orders of magnitude—but the onset of the high shear-rate limit is characterised by a Peclet number of order unity.

The dependence of the viscosity on volume fraction for model hard-sphere colloids is shown in Fig. 4.12 for both the high and low shear-rate limits. At the lowest volume fractions no shear-rate dependence is detectable, and the Einstein relation, eqn 4.48, is obeyed. At higher volume fractions, the high shear-rate relative viscosity can be fitted to the function

$$\frac{\eta_{\text{high}}}{\eta_0} = \left(1 - \frac{\phi}{0.71}\right)^{-2}. \tag{4.52}$$

At low shear rates a similar function

$$\frac{\eta_{\text{low}}}{\eta_0} = \left(1 - \frac{\phi}{0.63}\right)^{-2} \tag{4.53}$$

provides a good fit.

The most striking feature of the viscosities at both high and low shear-rate limits is that they appear to diverge at a finite volume fraction. In the low shear-rate limit, the volume fraction, 0.63, at which the viscosity appears to diverge corresponds quite closely to the volume fraction for random close packing of hard spheres. At slightly lower volume fractions than this we would expect this divergence in viscosity to lead to a glass transition, and indeed it is found that random hard-sphere colloids with volume fractions between about 0.58 and the random close-packing limit form long-lived metastable states, rather than ordering to the thermodynamically favoured, lower free energy, close-packed crystal. The higher volume fraction marking the divergence in the high shear-rate limit is closer to the volume fraction for a close-packed crystal; this must reflect the ordering induced by the shear flow at these higher shear rates.

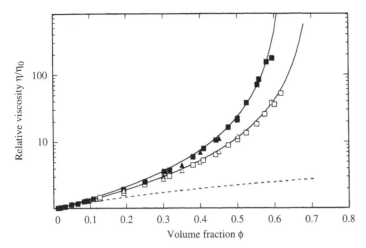

Fig. 4.12 Relative viscosity as a function of volume fraction for model hard-sphere latices, in the limit of low shear rates (filled symbols) and high shear rates (open symbols). The solid lines are the fitting functions, eqns 4.53 and 4.52, and the dashed line is the Einstein prediction for the dilute limit, eqn 4.48. Squares are 76 nm silica spheres in cyclohexane (de Kruif *et al.* 1986), triangles are polystyrene spheres of radii between 54 and 90 nm in water (Krieger 1972). After Russel *et al.* (1989).

Further reading

An excellent general introduction to the subject of intermolecular, interparticle, and intersurface forces is given in Israelachvili (1992). More details on the calculations can be found in Mahanty and Ninham (1976), from which the treatment of the Casimir effect given above was derived.

A good general overview of the subject of colloidal dispersions can be found in Russel *et al.* (1989). An interesting and accessible explanation of the relevance of colloid science to food can be found in Dickinson (1992).

For more details about the fractal aspects of colloidal aggregation, see Avnir (1989).

Exercises

(4.1) For each of the following particles, suspended in water,

    a) a grain of sand, $100\,\mu$m in diameter, density $2200\,\text{kg m}^{-3}$,

    b) a polymer particle, $1\,\mu$m in diameter, density $1050\,\text{kg m}^{-3}$,

    c) a virus particle, $50\,$nm in diameter, density $1020\,\text{kg m}^{-3}$,

        i. calculate the terminal velocity,

        ii. calculate the diffusion coefficient,

        iii. estimate the time taken for the particle to diffuse a distance equal to its own diameter.

[Data: Viscosity of water $= 1.002 \times 10^{-3}$ Pa s; density of water $= 1000\,\text{kg m}^{-3}$.]

(4.2) Consider a colloid consisting of particles of mass m and density ρ_c suspended in water.

    a) Show that the number density as a function of the height from the bottom of the container z, $n(z)$, is given by

$$n(z) = n_0 \exp\left(-\frac{m(\rho_c - \rho_w)gz}{\rho_c k_B T}\right), \quad (4.54)$$

where ρ_w is the density of water, g is the acceleration due to gravity, and n_0 is the number density at the bottom of the container.

b) Consider a slice of the suspension between heights z and $z + dz$.

 i. Show that diffusion leads to a net flux of particles J_D *into* the slice given by

$$J_D = D \frac{m^2 g^2}{k_B^2 T^2} \frac{(\rho_c - \rho_w)^2}{\rho_c^2} n(z)\, dz, \quad (4.55)$$

where D is the diffusion coefficient of the colloidal particles.

 ii. Show that sedimentation under gravity leads to a net flux of particles *out* of the slice J_s given by

$$J_s = \frac{m^2 g^2}{\xi k_B T} \frac{(\rho_c - \rho_w)^2}{\rho_c^2} n(z)\, dz, \quad (4.56)$$

where ξ is the drag coefficient.

 iii. Use the Stokes–Einstein law to show that the flux into the slice due to diffusion is balanced by the flux out of the slice due to sedimentation.

(4.3) Consider the van der Waals interaction between two semi-infinite gold plates interacting across a vacuum.

a) Compare the force per unit area as predicted by the Hamaker approach, taking the value of the Hamaker constant $A = 2 \times 10^{-19}$ J, with the value of the Casimir force, for values of the plate separation of

 i. $1\,\mu$m,
 ii. 100 nm,
 iii. 1 nm.

b) At what separation h_x are the two forces predicted to be equal?

c) Which expression is more likely to be accurate for separations greater than h_x, and which is more accurate for separations less than h_x? Give reasons for your answer in each case.

(4.4) Consider a colloid of charged spheres all of radius $0.1\,\mu$m in an aqueous solution of sodium chloride.

a) Calculate the Debye screening length for salt concentrations of 10^{-5}, 10^{-4}, 10^{-3}, and 10^{-2} mol/dm^3.

b) For each of the salt concentrations above, estimate the volume fraction for the transition to an ordered phase. You may assume that the particles may be considered to behave as hard spheres with an effective radius equal to the sum of the physical radius and the Debye screening length.

(4.5) A water-based varnish is composed of a dispersion of polymer spheres with diameters of 200 nm. If it is brushed onto a surface as a film of thickness $200\,\mu$m, how fast must the brush be moved to achieve an appreciable degree of shear thinning?

Polymers

5

5.1 Introduction

Polymers and polymer solutions and mixtures form a large and important class of soft condensed matter. Polymers may be synthetic in origin; plastics such as polystyrene and polyethylene are polymers, as are the components of many glues, fibres, and resins. Many materials of biological origin are also composed of polymers; examples of biologically important polymers include proteins, nucleic acids such as DNA, and polysaccharides like starch. Polymers, both natural and synthetic, are often major components in complex composite materials, both natural and artificial, such as glass-reinforced plastic, wood, and tissue.

Despite the very wide variety of different properties of polymers that arise from the different chemistry that makes them up, many of their physical properties have **universal** characteristics: these characteristics are a result of **generic** properties of long, string-like, molecules. An example of such a generic property is the fact that two molecules cannot cross one another; this leads to the effect of **entanglement**, which produces dramatic viscoelastic effects in polymer melts and solutions. Surprisingly, it proves possible to construct remarkably simple and general theories to account for these properties.

5.2 The variety of polymeric materials

A polymer is a giant molecule, a molecule made up of many repeat units covalently joined together in the form of a long chain. Chemists sometimes use a more circumscribed definition, restricting the word polymer to molecules in which every subunit is identical. This is too restrictive for our purposes. On the other hand, in other long-chain objects the subunits are joined not by covalent bonds, but by physical ones. Examples of this are the giant worm-like micelles formed in some amphiphile solutions (see Section 9.2.5), and the long chains of compact protein molecules which constitute, for example, actin filaments. Such objects are sometimes called 'living polymers'; their characteristic is that they can change their length in response to changes in the environment. This contrasts with the more usual covalently linked polymers, in which the length of the molecules, or the distribution of lengths, is fixed during the polymerisation process. In this chapter we restrict our attention to covalently linked polymers, but even here we find tremendous variety. In this section we will review this variety, before discussing those polymer properties that are more universal.

5.2.1   Polymer chemistry

Polymer chemistry is predominantly a branch of organic chemistry, in that most polymers are based on carbon. But even with this distinction there is a huge variety of possible structures, and from the different structures follow different properties. Some examples are shown in Fig. 5.1: the simplest chemical structure is a carbon main chain with two hydrogen atoms per carbon—**polyethylene**.

The same main chain can have different side groups, for example **poly(methyl methacrylate)**. The main chain can incorporate atoms which are not carbon, for example **nylon 6-10**. The main chain can involve loops, such as **amylose**, or

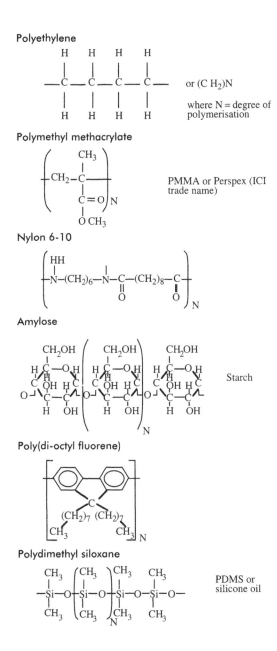

Fig. 5.1 Some different polymers.

it can be multiply connected, forming a ladder structure, such as **poly(dioctyl fluorene)**. Finally there are common polymers which do not involve carbon in their main chain at all, such as **poly(dimethyl siloxane)**.

In this book we make no mention at all of the many synthetic routes by which polymers are made—for this the reader is referred to a textbook on polymer synthesis. However, physicists should note with due humility the tremendous intellectual and practical achievements of polymer chemists, who have devised synthetic routes allowing precise control of polymer composition and architecture. Without these developments few of the advances in understanding made by polymer physics would have been possible.

5.2.2 Stereochemistry

If a polymer has more than one type of chemical group attached to each main chain carbon atom, then different arrangements of the groups in three dimensions are possible. This is illustrated in Fig. 5.2, which shows three possible arrangements of a simple carbon main chain polymer with two different groups per carbon atom. For example, if X is a hydrogen atom and Y is CH_3, this would be polypropylene.

There are two regular arrangements of side groups: these are called **isotactic** and **syndiotactic**. In these the similar side groups appear on the same side of the chain or on alternate sides respectively. If the arrangement of the groups is random, then we have an **atactic** polymer.

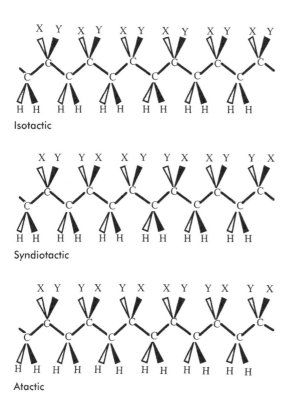

Fig. 5.2 Stereochemistry in polymers.

The importance of the atactic arrangement is that it involves **quenched disorder**. By this we mean that the energy barriers to rotation of the side groups are so large that once the arrangement is set in place during the synthesis of the polymer, no rearrangement of the groups can subsequently take place. Thus even at absolute zero, when the entropy of an equilibrium system should go to zero, there will still be some residual disorder. The quenched disorder inherent in atactic polymers means that these molecules usually find it impossible to crystallise—at low temperatures these materials form **glasses**.

5.2.3 Architecture

According to the way in which they are synthesised, chains may be **linear** or **branched**. Linear polymers are completely characterised by the number of repeat units present in the chain, the **degree of polymerisation**, N. This is proportional to the **relative molecular mass** of the chain, M. Examples of linear polymers include synthetic chains such as high-density polyethylene, as well as some biopolymers, for example most simple proteins.

Almost all synthetic polymers are made by processes that yield not a single degree of polymerisation, but a distribution. This distribution can be characterised by various averages, of which the most important are the weight average and the number average. The ratio of these two averages gives a measure of the broadness of the distribution of degrees of polymerisation, and is known as the **polydispersity index**. The polydispersity of synthetic polymers, particular those that are commercially available in large quantities, poses substantial difficulties in understanding their properties. Fortunately for the point of view of fundamental studies, chemists have been able to devise synthetic routes to obtain polymers with narrow distributions of degree of polymerisation.

Both synthetic and natural polymers may be branched. For example, low-density polyethylene contains many branches. Once again, these branched points introduce **quenched disorder**, and hinder the formation of crystals in these materials. The presence of branches also substantially changes the way in which molecules can move relative to one another, as we shall see in a later section, with profound effects on their rheology.

As more and more branch points are introduced to a polymer system, one can create a network that spans a macroscopic volume. In this case the polymer molecule essentially assumes macroscopic dimensions. Examples of such networks include vulcanised rubber and cured epoxy resins; the formation and properties of networks form the subject of Chapter 6.

5.2.4 Copolymers

In many polymers, the molecule is formed from a single type of repeat unit—such molecules are called **homopolymers**. However, other polymers are composed of more than one type of repeat unit. These molecules are called **copolymers**. Molecules with strikingly different properties are obtained according to the way in which the different repeat units are arranged.

A purely random arrangement of repeat units gives a **random copolymer**. Once again, the random distribution of groups which is locked in by covalent bonds gives a molecule with quenched disorder; this generally prevents these materials from crystallising well. Apart from this, the properties of such random copolymers are often found to be intermediate between the properties of the two homopolymers that would be formed from the two repeating units.

Dramatically new properties emerge if the different repeating units are arranged in blocks. In such a **block copolymer**, if the two (or more) types of repeating unit are very different chemically, they may have a tendency to phase-separate. But because the blocks are covalently linked, complete macroscopic phase separation is not possible. Instead, these materials **microphase separate** into a wide variety of possible complex morphologies. This tendency to self-assembly is discussed in Chapter 9.

A final, very special class of copolymer is formed by the biopolymers DNA and proteins. These are made up from repeating subunits of different chemical types which are arranged in a strictly prescribed, but non-periodic, sequence. Such **sequenced copolymers** are at first sight similar to random copolymers, but there is a profound difference. Whereas a synthesis that produces random copolymers will produce an ensemble of all possible arrangements of the repeating units, the synthesis of a sequenced copolymer produces just one arrangement—a sequence that has been selected by evolution to have a particular property. In the case of many proteins, this property is the ability of a single molecule to **self-organise** into a complex but well-defined three-dimensional structure. We discuss this further in Chapter 10.

5.2.5 Physical state

At a given temperature, and following a given processing route, polymeric materials can be found in a variety of physical states.

Liquid. Polymer melts and solutions are liquids, but they often are very viscous and show marked viscoelastic properties.

Glass. Because of the difficulty of crystallising polymers, polymer glasses are very common. Familiar examples include polystyrene and poly(methyl methacrylate).

Crystalline. Polymers can sometimes crystallise, but crystallisation is usually not complete, owing to kinetic limitations and the presence of quenched disorder. In such **semi-crystalline** materials, very small crystals exist in a matrix of amorphous material, which can be in either a liquid-like or glassy state. Familiar examples of semi-crystalline materials include synthetic polymers such as polyethylene as well as natural materials like starch. This state is discussed in more detail in Chapter 8.

Liquid crystalline. Some polymers are rather rigid molecules, which can line up to form liquid crystalline phases. These can form the basis of very strong engineering materials, like Kevlar. This state is discussed in Chapter 7.

5.3 Random walks and the dimensions of polymer chains

The simplest example of the way in which we can find universal behaviour amongst the variety of types of polymers occurs when we consider the overall dimensions of a polymer chain, and how these are related to its degree of polymerisation N. We find that an extremely simple model accounts quantitatively for many properties of polymer chains and provides the starting point for much of the physics of polymers. This model is that of an **ideal random walk**, which is realised in the idealisation of a **freely jointed chain**.

5.3.1   The freely jointed chain and its Gaussian limit

In this model we consider the chain to be made up of N links, each of which has length a. The different links have independent orientations. Thus the path of the polymer in space is a **random walk**.

The end-to-end vector $\mathbf{r}$ is simply the sum of the N jump vectors $\mathbf{a}_i$ which represent the direction and size of each link in the chain:

$$\mathbf{r} = \mathbf{a}_1 + \mathbf{a}_2 + \cdots + \mathbf{a}_N = \sum_{i=1}^{N} \mathbf{a}_i. \tag{5.1}$$

The mean end-to-end distance is

$$\langle \mathbf{r} \cdot \mathbf{r} \rangle = \left\langle \left(\sum_{i=1}^{N} \mathbf{a}_i \right) \cdot \left(\sum_{j=1}^{N} \mathbf{a}_j \right) \right\rangle. \tag{5.2}$$

Expanding the sum gives us

$$\langle \mathbf{r}^2 \rangle = \left\langle \sum_i \sum_j \mathbf{a}_i \cdot \mathbf{a}_j \right\rangle; \tag{5.3}$$

so picking out the N cases where $i = j$ we have

$$\langle \mathbf{r}^2 \rangle = Na^2 + \left\langle \sum_{i \neq j} \mathbf{a}_i \cdot \mathbf{a}_j \right\rangle. \tag{5.4}$$

If the chain is freely jointed, the directions of different links are completely uncorrelated and the cross-terms disappear when the average is taken. We recover the familiar random walk result

$$\langle \mathbf{r}^2 \rangle = Na^2. \tag{5.5}$$

The overall size of a random walk is proportional to the **square root** of the number of steps.

What is the distribution of possible end-to-end distances? This calculation is a relatively straightforward piece of statistical mechanics that is dealt with in Appendix A. The answer is that in the limit of large N this distribution is **Gaussian**. Specifically, the probability distribution function $P(\mathbf{r}, N)$ is given by

$$P(\mathbf{r}, N) = \left(\frac{2\pi Na^2}{3} \right)^{-3/2} \exp \left(-\frac{3\mathbf{r}^2}{2Na^2} \right). \tag{5.6}$$

This expression is very important in understanding the physics of polymers, because it allows us to write down the **configurational entropy** $S(\mathbf{r})$ of a polymer chain as a function of its elongation:

$$S(\mathbf{r}) = -\frac{3k_B\mathbf{r}^2}{2Na^2} + \text{constant.} \tag{5.7}$$

Thus when we stretch a polymer chain its entropy is lowered. This results in an **increase** in the free energy $F(\mathbf{r})$:

$$F(\mathbf{r}) = \frac{3k_{\mathrm{B}}T\mathbf{r}^2}{2Na^2} + \text{constant}. \tag{5.8}$$

A polymer chain behaves like a spring; if it is stretched beyond its ideal random walk value there is a restoring force proportional to the extension. However, this restoring force does not arise from an increase of internal energy of the polymer, as it would for a spring; instead the origin of the force is entirely **entropic**. There are fewer possible configurations of the polymer chain when it is stretched than when it is in its unperturbed, random walk state.

5.3.2   Real polymer chains—short-range correlations

The freely jointed chain model has the appearance of being rather unphysical. We know from elementary chemistry that successive links in a polymer chain are not free to rotate, but instead are constrained to have certain definite bond angles. Thus in the derivation given above, the cross-terms describing the correlations of neighbouring bonds do not disappear. The surprising and important result of more careful calculations is that the presence of these short-range correlations does not alter the basic random walk character of the polymer chain statistics; it simply leads to an altered effective step size.

Consider, for example, a model in which the bond is free to rotate, but has a definite bond angle θ. Now, if we redo the calculation of the last section we find for the cross-terms of eqn 5.4, $\langle \mathbf{a}_i \cdot \mathbf{a}_{i-1} \rangle = a^2 \cos \theta$, and in general $\langle \mathbf{a}_i \cdot \mathbf{a}_{i-m} \rangle = a^2 \cos^m \theta$. As $\cos \theta$ is always less than unity, the correlation dies away along the chain. This means that we can conceive of breaking the chain up into subunits, whose size we choose to be larger than the range of the correlations.

Suppose there are g links in one of these new subunits, whose vectors are written $\mathbf{c}_i$. Then the end-to-end distance is given by

$$\langle \mathbf{r}^2 \rangle = \frac{N}{g} \langle \mathbf{c}^2 \rangle = Nb^2, \tag{5.9}$$

where b is an effective monomer size, the statistical step length. In practice the effect of correlations along the chain is often characterised in terms of the characteristic ratio C_∞:

$$C_\infty = \frac{b^2}{a^2}. \tag{5.10}$$

If one knows the chemical details of the polymer the characteristic ratio or the statistical step length may be calculated; alternatively one can extract their values for a given polymer from experimental data on chain dimensions.

This simple example is typical of the way in which polymer physics tries to proceed: the long-range structure (in this case the scaling of chain dimensions with the square root of the degree of polymerisation N) is given by statistics and is **universal** (i.e. independent of the chemical details of the polymer in question). All these chemical details go into one parameter, which may be either calculated on the basis of detailed theory at the atomic level, or extracted from experiment.

5.3.3 Excluded volume, the theta temperature, and coil–globule transitions

In the argument above for the mean end-to-end distance of a polymer we accounted for interactions between neighbouring links of the chain, but we neglected interactions between distant points on the chain. It turns out that for an isolated chain these interactions are important, and affect the long-range structure of the chain.

At the simplest level, we know that the chain cannot intersect itself; the monomer units have finite volume and two cannot be in the same place at the same time. We have a self-avoiding walk. The mathematics of self-avoiding walks is much more complicated than for simple random walks; instead of finding $\langle \mathbf{r}^2 \rangle^{1/2} = a\sqrt{N}$, the random walk result, we find $\langle \mathbf{r}^2 \rangle^{1/2} = aN^{\nu}$, where the exponent $\nu > 0.5$. The effect of excluded volume, then, is to swell the polymer chain over the random walk value.

What is the value of the exponent ν? We can estimate it using a very simple argument due to Flory. We consider the polymer molecule to be a gas of N segments confined to a volume r^3, where r is the radius of the polymer coil that we need to determine. The concentration of segments, c, then, is

$$c \sim \frac{N}{r^3}. \tag{5.11}$$

We introduced the idea of excluded volume in Section 4.4.1 in the context of the crystallisation of hard-sphere colloids. There we saw that in a gas of N atoms confined to a volume V, the entropy per atom is reduced by a factor of $k_{\mathrm{B}} v N / V$ if each atom occupies a volume v which is thus inaccessible to the other atoms. Thus in our polymer coil there is a positive contribution to the free energy due to the excluded volume F_{rep} which we can write as

$$F_{\mathrm{rep}} = k_{\mathrm{B}} T v \frac{N^2}{2r^3}. \tag{5.12}$$

So one can think of the segments within the chain repelling each other as a result of the excluded volume effect, which tends to make the chain expand. But this expansion is opposed by the effect of configurational entropy—the entropic spring effect that we introduced in Section 5.3.1. This gives an elastic contribution to the free energy F_{el}

$$F_{\mathrm{el}} = k_{\mathrm{B}} T \frac{r^2}{Na^2}. \tag{5.13}$$

If we now write down the total free energy $F_{\mathrm{total}} = F_{\mathrm{rep}} + F_{\mathrm{el}}$ and minimise it with respect to r, we find

$$r \sim aN^{3/5}. \tag{5.14}$$

(Here we have assumed that the excluded volume $v \sim a^3$.)

Thus Flory's argument suggests that the exponent $\nu = 3/5$. This is very close to the actual value, which can be computed using complicated renormalisation group techniques and is (to three decimal places) 0.588.

Experimentally, the chain dimensions of very dilute chains in good solvents are found to be swollen in the way predicted (though available experimental precision is not really able to distinguish between the Flory value of 3/5 and the more exact value).

In the argument just given, we have neglected any energetic interaction between neighbouring polymer segments or between polymer segments and neighbouring solvent molecules. We can introduce such interactions in a simple way similar to the method introduced to deal with interactions in liquid mixtures in Section 4.2. There we assumed we could calculate such energies just by adding up all the pairwise interactions between neighbouring molecules or segments. So we take the energy of interaction between two neighbouring polymer segments as ϵ_{pp}, between two neighbouring solvent segments as ϵ_{ss}, and between a neighbouring polymer and solvent segment as ϵ_{ps}. If we assume that the polymer segments are uniformly distributed within the coil with a concentration c given by eqn 5.11, then we can estimate the number of contacts between different polymer segments N_{pp}, between solvent molecules N_{ss}, and between polymer segments and solvent molecules N_{ps}, as

$$N_{pp} = \frac{1}{2}zNvc$$
$$N_{ps} = zNv(1 - c) \tag{5.15}$$
$$N_{ss} = N_{ss}^0 - \frac{1}{2}zNvc - zNv(1 - c),$$

where N_{ss}^0 is the number of solvent/solvent contacts there would be in the absence of a polymer chain, and z is the number of neighbours possessed by each solvent molecule or polymer segment.

Adding up the total energy of interaction U_{int} we find

$$U_{int} = \frac{1}{2}zNvc(\epsilon_{pp} + \epsilon_{ss} - 2\epsilon_{ps}) + zN(\epsilon_{ps} + \epsilon_{ss}) + N_{ss}^0\epsilon_{ss}. \tag{5.16}$$

As we did in Section 4.2, we can combine the interaction energies into a single interaction parameter χ, given by

$$\chi k_B T = \frac{1}{2}z(2\epsilon_{ps} - \epsilon_{pp} - \epsilon_{ss}), \tag{5.17}$$

and then write the interaction energy as a function of the chain size r as

$$U_{int} = -k_B T 2v\chi \frac{N^2}{2r^3} + \text{constant}. \tag{5.18}$$

This equation has exactly the same functional form as eqn 5.12 for the repulsive free energy due to excluded volume, so we can write down an expression for the combined excluded volume and solvent interaction energies as

$$F_{rep} + U_{int} = k_B T v(1 - 2\chi)\frac{N^2}{2r^3} + \text{constant}. \tag{5.19}$$

Thus, according to the value of χ, we have three different kinds of behaviour:

1. $\chi < 1/2$: the polymer chain is expanded, with a radius $r \sim N^{3/5}$. This is referred to as **good solvent** behaviour.
2. $\chi = 1/2$: the repulsive effect of excluded volume is exactly cancelled by the attractive effect of the polymer/solvent interactions. The polymer chain is an ideal random walk, with $r \sim N^{1/2}$. This situation is known as the **theta condition**.

3. $\chi > 1/2$: the attractive effect of the polymer/solvent interaction out-weighs the repulsive excluded volume interaction, and the chain collapses to form a compact **globule**.

As the value of the interaction parameter χ between a polymer and a solvent is changed, for example by changing the temperature, the polymer undergoes a transition from an expanded chain, or coil, through the theta temperature to a collapsed coil. This is known as the **coil-globule** transition. Our treatment of excluded volume and the interaction energy is too crude to allow us to deduce the details of the shape of the transition, but a more sophisticated treatment reveals that the transition is a thermodynamic phase transition in the limit of long chains, $N \to \infty$; the transition may be first or second order according to the stiffness of the chains.

5.3.4 Chain statistics in polymer melts—the Flory theorem

In dilute solution, polymers do not follow random walk statistics; because of excluded volume interactions, the chain is swollen.

What happens as we make the solution more concentrated, ultimately arriving at the polymer melt? Counterintuitively, rather than becoming more complicated, things become more simple: in the melt chains follow ideal random walk statistics!

This was first realised in 1949 by Flory and put on a sound theoretical footing by Edwards in 1966, when he introduced the idea of screening. Here is a simple plausibility argument.

Consider schematic plots of the segment concentration c across a section through space (Fig. 5.3; for a dilute solution chains are isolated and segments do not interact with segments from different chains).

The effect of the excluded volume interaction is to include an unfavourable energy proportional to the probability of two segments being close together; that is, proportional to c^2. This leads to a force on the segments proportional to $c \, dc/dx$, which tends to make the chain expand. Now as the concentration is increased chains start to overlap; dc/dx is smaller and the repulsive force is reduced.

Finally, in the melt, the concentration of segments is essentially uniform; thus there is no repulsive force between the segments and the chain follows ideal random walk statistics. This result was confirmed by neutron scattering in the early 1970s.

5.3.5 Measuring the size of polymer chains

The dimensions of polymer chains are most directly measured by scattering. This method relies on the interference of waves scattered from different parts of an extended scattering object.

At large angles, waves scattered from different parts of the object interfere destructively. The way the intensity drops off with angle depends on the relative magnitude of the inverse size of the object and the scattering vector Q, which is related to the wavelength of the wave λ and the angle through which it is scattered θ by the relation $Q = 4\pi \sin\theta/\lambda$.

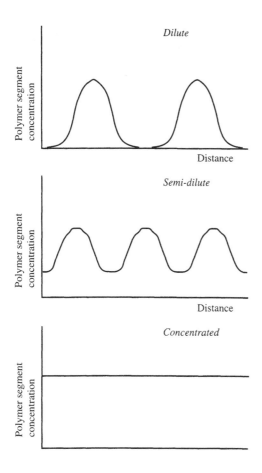

Fig. 5.3 Screening in polymer solutions. In a dilute solution, the local concentration of polymer segments through a section of the solution shows strong peaks corresponding to the location of individual molecules; this results in an outward force arising from osmotic pressure which tends to swell the individual chains. As the concentration of solution is increased, chains increasingly overlap; the variations in local concentration become less pronounced and the force driving the swelling of individual coils is reduced. In concentrated solutions or melts, the local concentration of segments is uniform and there is no force tending to swell the chains.

The size of the object is characterised by the **radius of gyration** R_g; formally this is defined as the mean squared distance of each point on the object from its centre of gravity. For a random walk it can be shown that $R_g^2 = \langle R_{\text{end-to-end}}^2 \rangle /6$.

In deciding what kind of radiation to use, we have to consider two factors:

wavelength—the radiation must have a wavelength comparable to the size of the object, so that we can detect scattering from the object;

contrast—the radiation must interact differently with the object than with the medium it is immersed in.

The possible types of radiation are as follows:

Visible light—here contrast is based on refractive index differences between the scattering object and its environment. The relatively long wavelength compared to molecular sizes means that even quite large polymers scatter to wide angles.

X-rays—here the contrast is based on differences in electron density between the scattering object and its environment. In practice this means that the scattering object must contain atoms of relatively large atomic number compared to the surroundings. The smaller wavelength than light means that objects with sizes typical of polymer molecules scatter at small angles.

Neutrons—here the contrast is based on differences in nuclear interactions. As it happens, the scattering of neutrons is very different for hydrogen and deuterium, allowing one to distinguish a polymer molecule from a chemically very similar environment by **deuterium labelling**. The wavelength of thermal neutrons is similar to X-rays.

Although neutron scattering is expensive, the fact that it can distinguish between hydrogen and deuterium means that it can measure chain dimensions in concentrated solutions and melts—this cannot be done any other way.

5.3.6 Polymers at interfaces—adsorbed and grafted chains

We have already seen in Section 4.3.5 that polymer chains bound to surfaces can profoundly modify the interactions between such surfaces, giving a mechanism for the stabilisation of colloidal dispersions. We are now in a position to understand the physics underlying this steric stabilisation in more detail.

Two important ways in which polymers can interact with a surface are illustrated in Fig. 5.4. In the case of an adsorbed homopolymer, relatively weak energetically favourable interactions between an individual segment and the surface combine to anchor the chain firmly to the surface. The conformation of the polymer chain must be perturbed by its interaction with the surface, and we would expect more severely perturbed conformations—corresponding to a more flattened arrangement of the chain—for higher energies of interaction. However, it is rather difficult to calculate the precise relationship between the adsorption energy and the overall conformation of the whole chain.

The situation in which chains are attached by one end to an interface turns out to be much more easily dealt with theoretically, as well as being relevant to a number of practical situations. For a relatively densely grafted layer of polymers—often referred to in the literature as a 'polymer brush'—we can calculate the height of the brush using the kind of approach we introduced in Section 5.3.3. Suppose we have polymers of degree of polymerisation N grafted to a surface at a density of σ/a^2 chains per unit area. The volume per chain is ha^2/σ. We can write the total energy as the sum of a stretching energy

$$F_{\text{el}} = k_{\text{B}}T\frac{h^2}{Na^2} \tag{5.20}$$

per chain, and a combined excluded volume and interaction energy

$$F_{\text{rep}} + U_{\text{int}} = k_{\text{B}}Tb(1 - 2\chi)\frac{\sigma N^2}{2ha^2}. \tag{5.21}$$

Fig. 5.4 Polymers at interfaces. (a) Adsorbed polymers. A positive interaction between segments along the chain and the interface means that the polymer is adsorbed at a number of points along the chain, with loops and tails of polymer protruding into the solvent. (b) End-anchored polymers. The polymers are physically or chemically attached by one end; the chains protrude into the solvent forming a **polymer brush**.

(a)

(b)

Finding the value of the brush height h that minimises the total energy gives us

$$h \sim [\sigma b(1 - 2\chi)]^{1/3}N. \qquad (5.22)$$

This reveals that a chain end-grafted to a polymer surface at high enough densities to overlap significantly with its neighbours is **strongly stretched**— the chain dimensions vary not as $N^{3/5}$, as they do for an isolated chain in a good solvent, but as a linear function of N.

5.4 Rubber elasticity

The first statistical mechanical theory of the mechanical properties of a polymeric material was the classical theory of rubber elasticity.

A **rubber** is a polymer melt in which cross-links, randomly placed between adjacent chains, bond the chains together to form a macroscopic network.[1] At a local level, the material behaves like a liquid; in particular the bulk modulus is rather high and to a first approximation the material may be taken as incompressible. However, the cross-links mean that macroscopic bulk flow cannot take place, and the material has a finite shear modulus. The classic theory of rubber elasticity uses statistical mechanics to calculate the shear modulus.

[1] In a classical rubber, the chains are linked by covalent bonds in the process of **vulcanisation**. However, a variety of physical and chemical mechanisms can result in cross-links, some of which are reviewed in Chapter 6.

To calculate the modulus, we start by making an important assumption: when the rubber is deformed each individual cross-link point moves in proportion to the deformation of the whole sample. This is the assumption of **affine deformation**. Thus if the initial dimensions of the sample are l_x, l_y, and l_z, and after deformation these dimensions become $\lambda_x l_x$, $\lambda_y l_y$, and $\lambda_z l_z$, the position of a point at coordinates (x, y, z) transforms to $(\lambda_x x, \lambda_y y, \lambda_z z)$.

If we consider a single strand between two cross-link points, one of which is initially at the origin, and the other is at the coordinates (x, y, z), we find its initial mean end-to-end distance $r_0^2 = x^2 + y^2 + z^2$. After deformation, the new mean end-to-end distance $r_1^2 = \lambda_x^2 x^2 + \lambda_y^2 y^2 + \lambda_z^2 z^2$. Using the relation for the entropy of a Gaussian chain given in Section 5.3.1, we find the change in entropy per strand on deformation, ΔS_{strand}, to be given by

$$\Delta S_{\text{strand}} = \frac{-3k_B}{2Na^2}\left[(\lambda_x^2 - 1)x^2 + (\lambda_y^2 - 1)y^2 + (\lambda_z^2 - 1)z^2\right]. \qquad (5.23)$$

If there are n strands per unit volume, we can use the relation $\langle x^2 \rangle = \langle y^2 \rangle = \langle z^2 \rangle = Na^2/3$ to write the total entropy change per unit volume ΔS_{volume} that results from the deformation as

$$\Delta S_{\text{volume}} = \frac{-nk_B}{2}(\lambda_x^2 + \lambda_y^2 + \lambda_z^2 - 3). \qquad (5.24)$$

If we take the simplest case of deformation, a simple elongation in the x direction, for which, if we assume that rubber is incompressible, $\lambda_x = \lambda$, $\lambda_y = 1/\sqrt{\lambda}$, and $\lambda_z = 1/\sqrt{\lambda}$,[2] we find for the entropy change

[2] Because the incompressibility condition implies that $\lambda_x \lambda_y \lambda_z = 1$.

$$\Delta S_{\text{volume}} = \frac{-nk_B}{2}\left(\lambda^2 + \frac{2}{\lambda} - 3\right). \qquad (5.25)$$

Associated with this change in entropy is a change in free energy ΔF_{volume} given by

$$\Delta F_{\text{volume}} = \frac{-nk_B T}{2}\left(\lambda^2 + \frac{2}{\lambda} - 3\right). \qquad (5.26)$$

This is simply the elastic work done during the deformation. From this we can obtain the relation between tensile stress τ and strain e by noting that $\lambda = 1 + e$, and $\tau = dF/de$. Thus

$$\tau = nk_{\mathrm{B}}T \left[(1+e) - \frac{1}{(1+e)^2} \right]. \tag{5.27}$$

This is a non-Hookean stress/strain behaviour, but we can expand for small strain to find that the Young modulus E is $E = 3nk_{\mathrm{B}}T$. Once again assuming incompressibility (and thus a Poisson ratio of 0.5) this gives for the shear modulus G the very simple relation

$$G = nk_{\mathrm{B}}T. \tag{5.28}$$

Another convenient way to express this result, which we will use later, is in terms of M_x, the average relative molecular mass between cross-links, and the density ρ:

$$G = \frac{\rho RT}{M_x}, \tag{5.29}$$

where R is the gas constant.

This remarkable result indicates that the mechanical properties of a rubber are functions only of the temperature and cross-link density; as long as the chains behave in a liquid-like way (i.e. we are not in a glass) the chemistry of the polymer chains and the nature of the cross-links are not relevant.

5.5 Viscoelasticity in polymers and the reptation model

Polymer melts are usually highly viscous, but they also show very marked viscoelastic behaviour. In this section we will review the basic phenomenology of polymer viscoelasticity, before introducing a powerful and simple model that allows us to understand and make quantitative predictions of polymer viscoelasticity from a molecular point of view. This model is the theory of **reptation**.

5.5.1 Characterising the viscoelastic behaviour of polymers

We saw in Chapter 2 how viscoelastic liquids can behave either elastically or viscously, depending on the timescale over which they are deformed. Here we define measured functions that can characterise the observed viscoelastic properties. In each case, the function is defined with reference to an idealised experiment.

The **creep compliance** $J(t)$ is defined with reference to an experiment in which a stress σ_0 is applied at time $t = 0$ and held constant, and the strain $e(t)$ is followed as a function of time. The creep compliance is then given by

$$e(t) = \sigma_0 J(t). \tag{5.30}$$

(a)

(b)

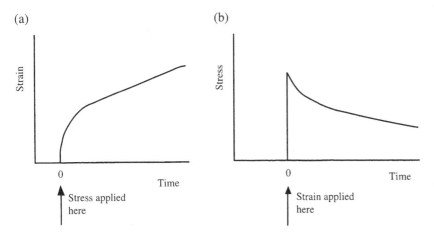

Fig. 5.5 (a) The creep compliance. A stress is applied at time $t = 0$ and held constant. The strain e is followed as a function of time. (b) The stress relaxation modulus. A strain is applied at time $t = 0$ and held constant. The stress σ is followed as a function of time.

The form of this function for a typical viscoelastic material is sketched in Fig. 5.5(a). After an initial fast, elastic response, the sample slowly creeps before settling down to long-term viscous behaviour in which the strain rate is constant.

The **stress relaxation modulus** $G(t)$ is defined by an experiment in which a strain e_0 is applied at time $t = 0$ and held constant, and the stress $\sigma(t)$ followed as a function of time. The stress relaxation modulus is then given by

$$\sigma(t) = e_0 G(t). \tag{5.31}$$

A typical shape for the stress relaxation modulus is sketched in Fig. 5.5(b). The initial response is elastic; as the material starts to flow the stress falls away to zero.

An extremely important type of deformation, both experimentally and for theoretical considerations, is an oscillatory deformation at a certain frequency ω. If an oscillatory strain of the form

$$e(t) = e_0 \cos(\omega t) \tag{5.32}$$

is applied, then (for small amplitudes) the resulting stress can be written as

$$\sigma(t) = e_0 \left[G'(\omega) \cos(\omega t) - G''(\omega) \sin(\omega t) \right] \tag{5.33}$$

where $G^*(\omega) = G'(\omega) + G''(\omega)$ is the **complex modulus**. The complex modulus and the stress relaxation modulus are related by the expression

$$G^*(\omega) = i\omega \int_0^\infty \exp(-i\omega t) G(t) \, dt. \tag{5.34}$$

The complex modulus G^* thus tells us about both the elastic response and the viscous response at any given frequency. The real part, G', is the elastic component of the response, and is known as the **storage modulus**. The imaginary part, G'', is the viscous component of the response and is known as the **loss modulus**. The phase angle of the response, δ, is given by $\tan \delta = G'/G''$.

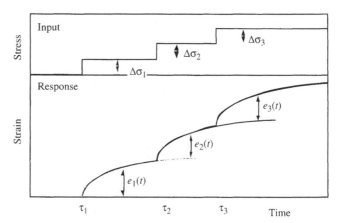

Fig. 5.6 The Boltzmann superposition principle. For small deformations, each loading step makes an independent contribution to the total loading history.

5.5.2   Linear viscoelasticity and the Boltzmann superposition principle

For small deformations one can assume that each loading step makes an independent contribution to the total loading history, and the final deformation is simply the sum of the response to each step, as illustrated in Fig. 5.6.

Thus it follows that we can write the strain as a function of time as an integral involving the creep compliance $J(t)$

$$e(t) = \int_{-\infty}^{t} J(t-\tau)\,d\sigma(t) = \int_{-\infty}^{t} J(t-\tau)\frac{d\sigma(t)}{d\tau}\,d\tau \qquad (5.35)$$

and, similarly, the stress as a function of time as an integral involving the stress relaxation modulus $G(t)$

$$\sigma(t) = \int_{-\infty}^{t} G(t-\tau)\frac{de(t)}{d\tau}\,d\tau. \qquad (5.36)$$

In the special case where the shear rate is constant this recovers Newton's law of viscosity $\sigma(t) = \eta_0 \dot{e}$ and gives us a useful relationship between the viscosity η_0 and the stress relaxation modulus:

$$\eta_0 = \int_{0}^{\infty} G(t-\tau)\,d\tau. \qquad (5.37)$$

The viscosity η_0 is called the **zero shear viscosity** to emphasise the fact that this relation relies on the assumptions of linear viscoelasticity. In fact it is very easy to go beyond the linear regime in experiments, but the theory of non-linear viscoelasticity is extremely complex and beyond the scope of this book.

5.5.3   The temperature dependence of viscoelastic properties: time–temperature superposition

We have seen that viscoelastic quantities are intrinsically a function of **time**. Experimentally, one also finds that the viscoelastic properties of polymers are a strong function of **temperature**; for example, a very viscous polymer solution or melt will become less viscous on being warmed. It turns out experimentally that the dependences of any viscoelastic quantity on time (or equivalently frequency) and temperature can be factorised. Thus, taking the stress relaxation modulus as a function of time t and temperature T, $G(t, T)$, as an example, we can write

$$G(t, T) = G(a_T t, T_0), \tag{5.38}$$

where $G(t, T_0)$ is measured at a reference temperature T_0 and a_T is a temperature-dependent **shift factor**. Empirically it is found that the temperature dependence of the shift factor can be written

$$\log a_T = \frac{-C_1(T - T_0)}{C_2 + (T - T_0)}. \tag{5.39}$$

It is common to use the glass transition temperature as the reference temperature T_0, in which case the constants C_1 and C_2 take values which are rather similar for many different polymers. Taking the logarithm as being to base 10, as is usual in the polymer literature, these quasi-universal values are $C_1^g = 17.4$ and $C_2^g = 51.6$ K. This expression is known as the WLF equation, after Williams, Landel, and Ferry.

The fundamental significance of time–temperature superposition is that it tells us that in the problem of viscoelasticity all the microscopic timescales scale in the same way with temperature. We can anticipate that in any microscopic theory of viscoelasticity the temperature will enter via a single temperature-dependent parameter. The specific form of the WLF relation for the shift factor given in eqn 5.39 can be shown to be equivalent to the statement that polymers have dynamics obeying the Vogel–Fulcher law, like most other glass-forming liquids (see Section 2.4.2).

In addition to its theoretical significance, the principle of time–temperature superposition is very useful experimentally, as it allows us to build up data sets covering very wide effective ranges of timescales on instruments whose actual range of accessible times is much smaller.

5.5.4 Viscoelasticity: experimental results for monodisperse linear polymer melts

Viscoelasticity is best understood for melts of linear polymers in which the distribution of relative molecular masses is rather narrow. For such materials the quantities characterising linear viscoelasticity show a strong universal character.

In Fig. 5.7, the stress relaxation modulus $G(t)$ is shown for two chemically identical polymers with different degrees of polymerisation. At short times, the curve is independent of degree of polymerisation. At intermediate times there is a wide range of times for which the modulus is essentially constant—this is the **plateau modulus**. The value of the plateau modulus does not depend on the degree of polymerisation. The plateau ends at a **terminal time** τ_T, which does depend strongly on degree of polymerisation, according to a power law $\tau_T \sim N^m$, where the exponent $m \approx 3.4$.

In the plateau part of the curve, the polymer behaves like an elastic solid. In fact the numerical value of the modulus is comparable to that of a lightly cross-linked rubber. After the terminal time the polymer flows in an essentially viscous manner; here the creep compliance $J(t) \sim t$. The zero-shear viscosity η_0 is related to the terminal time and the plateau modulus G_p as $\eta_0 \sim \tau_T G_p$.

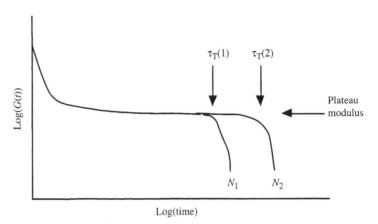

Fig. 5.7 Schematic diagram of the stress relaxation modulus $G(t)$ for monodisperse linear polymer melts. Curves are shown for two chemically identical polymers with different degrees of polymerisation N_1 and N_2, where $N_1 < N_2$.

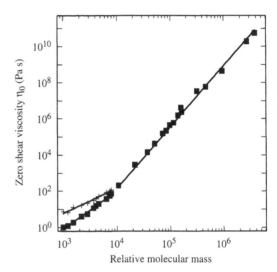

Fig. 5.8 Zero shear viscosity of polybutadiene as a function of the relative molecular mass. At high relative molecular mass, the viscosity fits a power law $\eta_0 \sim N^{3.4}$. At lower relative molecular mass, if the data is corrected for the relative molecular mass dependence of the glass transition temperature (crosses) we find $\eta_0 \sim N$. Data from Colby *et al.* (1987).

As the plateau modulus is independent of degree of polymerisation this shows us that the molecular weight dependence of the viscosity should be the same as that of the terminal time. Thus we expect $\eta_0 \sim N^m$, where the exponent $m \approx 3.4$.

This dependence of viscosity on degree of polymerisation is exactly what is observed whenever the degree of polymerisation is above a certain critical value N_c. This is illustrated in Fig. 5.8, which shows the dependence of zero shear viscosity on relative molecular mass for polybutadiene. Very similar curves are found for all linear polymers; below N_c the viscosity is proportional to N, while above N_c we find $\eta_0 \sim N^{3.4}$.

5.5.5 Entanglements

The striking features of polymer viscoelasticity can be understood at a qualitative level by the intuitively appealing notion of **entanglement**. Polymer chains are linear objects which cannot pass through one another; it is easy to imagine that in a melt of extended, interpenetrating chains an attempt to move the chains with respect to one another, by shearing the melt, will lead to the chains becoming tangled up with each other, making it difficult to shear the melt any further.

In this picture, one can interpret the rubber-like plateau in the shear modulus as indicating that entanglements behave like temporary cross-links. We can use the theory of rubber elasticity to estimate the density of entanglements, or equivalently the average distance along the chain between entanglements. As the plateau modulus is independent of degree of polymerisation, this quantity is a material constant for a given polymer. Adapting eqn 5.29, we find

$$G = \frac{\rho R T}{M_e},\qquad(5.40)$$

where M_e is the average relative molecular mass between entanglements, and the density is ρ.

These entanglements are not permanent—we imagine that they can disentangle with a characteristic timescale comparable to the terminal time τ_T. It is reasonable at a qualitative level that long chains will take longer to disentangle than shorter chains, but can we make this idea quantitative enough to explain the experimental result that $\tau_T \sim N^{3.4}$? To make progress towards a quantitative theory of viscoelasticity we need to make the idea of entanglements slightly more well defined. This has been achieved in the **tube model** and the idea of reptation, due to Edwards, de Gennes, and Doi.

5.5.6 The tube model and the theory of reptation

How is the motion of one chain constrained by the presence of all the other chains with which it is strongly entangled? The basic constraint is that no chains can cross each other's path. The effect of this is that each chain can be considered to be restricted to a **tube**. The motion of a chain in a tube is severely constrained laterally, but it is still free to wriggle along the length of the tube, like a snake moving through long grass (Fig. 5.9). This type of motion is called **reptation**.

We can now associate the terminal time with the time it takes the polymer to move completely out of its original tube. We can estimate this by assuming that within the tube motion is purely viscous, so that if we were able to move a segment of the chain through the tube it would feel a resistance force proportional to its velocity. Thus we can define a segment mobility μ_{seg}. If we were to pull a chain with N segments through the tube we would expect the viscous resistance felt by each segment to be additive, so mobility of the whole chain would be simply $\mu_{tube} = \mu_{seg}/N$.

Because the chain is an object in a viscous fluid, it will undergo constant Brownian motion; the chain will randomly wriggle backwards and forwards within the tube until it has completely escaped. Using Einstein's relation (see Section 4.2.2) we can write down a diffusion coefficient D_{tube} to describe this Brownian motion:

$$D_{tube} = k_B T \mu_{tube} = \frac{k_B T \mu_{seg}}{N}.\qquad(5.41)$$

Fig. 5.9 The tube model: the crosses (top) represent chains coming out of the plane of the paper. The test chain cannot cross these chains, and so it is confined to move in an effective tube (bottom).

The chain is doing what is essentially a one-dimensional random walk within the tube, so if the length of the tube is L, then the time taken to escape the tube, which we now identify with the terminal time τ_T, is given by

$$\tau_T = \frac{L^2}{D_{\text{tube}}}. \tag{5.42}$$

Since the length of the tube is proportional to the degree of polymerisation N, we thus find that

$$\tau_T \sim N^3. \tag{5.43}$$

This is quite close to the experimental result of $\tau_T \sim N^{3.4}$.

We can also derive another testable prediction from this theory, for the variation with degree of polymerisation of the **self-diffusion coefficient** of a polymer chain in a polymer melt. Each chain has diffused a distance comparable to its own end-to-end distance during the terminal time. Using the random walk result for chain end-to-end distance, we find

$$D_{\text{self}} \sim \frac{Na^2}{\tau_T}, \tag{5.44}$$

and thus

$$D_{\text{self}} \sim N^{-2}. \tag{5.45}$$

The experimental situation is illustrated in Fig. 5.10. Self-diffusion coefficients do obey a power law relation to the relative molecular mass, and for

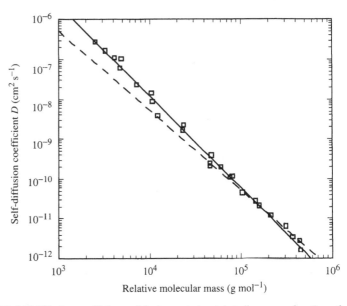

Fig. 5.10 Self-diffusion coefficients of hydrogenated polybutadiene as a function of relative molecular mass. The dashed line is the prediction of the simple reptation theory that $D_{\text{self}} \sim N^{-2}$, while the solid line is the best-fit power law, $D_{\text{self}} \sim N^{-2.30}$. Data from a number of authors, cited in Lodge (1999).

a number of years the experimental data was believed to be in good agreement with eqn 5.45. A more recent re-evaluation of the data (Lodge 1999) suggests that the exponent is slightly higher than 2; in fact the data seems to be best fitted by $D_{\text{self}} \sim N^{-2.30}$.

5.5.7 Modifications to reptation theory

Thus experimental data for viscosity, terminal time, and self-diffusion seems to be reasonably consistent with a single scaling of the terminal time with degree of polymerisation of $\tau_{\text{T}} \sim N^{(3.3-3.4)}$, while the simple reptation theory predicts $\tau_{\text{T}} \sim N^3$. What is the cause of this discrepancy?

It turns out that the reptation prediction must be modified by the inclusion of two other ways in which the tube constraint may be relaxed. These are:

1. **Constraint release.** In the tube model, it is assumed that the constraints that are described by the tube are immobile. In fact these constraints are themselves chains which are also reptating, so the tube must have a finite lifetime.
2. **Contour length fluctuation.** The length of the tube is defined, in effect, by the end-to-end distance of the polymer chain within it. This end-to-end distance **on average** is given by $a\sqrt{N}$, but as the chain undergoes Brownian motion this end-to-end distance will fluctuate. We can imagine the chain withdrawing into the tube, and then when it expands back to its average size some of the original constraints on it will have relaxed.

It is now believed that incorporation of these two corrections—particularly contour length fluctuations—can account for the discrepancy between experiment and the reptation prediction for the scaling of terminal time molecular weight. But perhaps the most striking implication of these effects comes from a consideration of the dynamics of **non-linear** polymers such as **star** and **branch** polymers.

The importance of these non-reptative relaxation modes for non-linear polymers is illustrated in Fig. 5.11. A star molecule is confined to a non-linear tube from which it cannot escape by reptation. Instead, if one of the arms is retracted into its tube by a contour length fluctuation the tube can relax. For this type of motion, one expects a relaxation time that is an **exponential** function of the arm length N_{arm}, rather than the power law dependence of a linear molecule.

This emphasises the importance of **architecture** in polymer physics. A linear polymer and a set of branched polymers of similar overall relative molecular mass, but with different degrees of branching, are chemically almost identical, but their physical properties, such as their melt viscosities, will be very different.

Further reading

There are a number of good introductions to polymer science generally, including some of the polymer chemistry that has been omitted in our treatment, for example Cowie (1998). Strobl (1997) is an excellent textbook which is focused on polymer physics.

There are three excellent theoretical monographs describing various aspects of polymer physics. De Gennes (1979) gives a lucid account of the applications

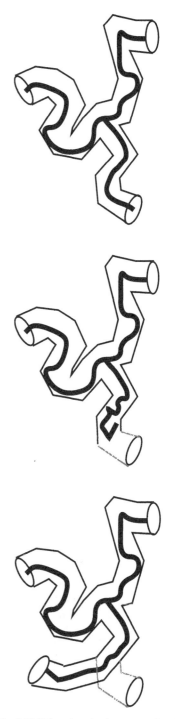

Fig. 5.11 Tube relaxation by contour length fluctuation in a star polymer. The star is unable to escape from its non-linear tube by reptation (top), but if one of the arms is retracted into the tube by a fluctuation in contour length (middle) it will re-emerge in a new relaxed configuration (bottom).

of scaling concepts to polymers, while Doi and Edwards (1988) gives the definitive account of the theory of reptation. Doi (1995) gives a concise introduction to some of these concepts. Grosberg and Khokhlov (1994) covers some similar ground to the other two monographs but also includes material on topics such as the coil–globule transition and the application of polymer theory to biopolymers. The same authors have also written a non-technical account of polymer science (Grosberg and Khokhlov 1997) which is well worth reading.

Exercises

(5.1) For polystyrene chains with a degree of polymerisation 10^4, calculate

    a) the RMS end-to-end distance in a melt,
    b) the RMS end-to-end distance in a dilute, good solvent, with a value of the interaction parameter $\chi = 0$.

[Take the monomer size to be equal to the statistical step length $a = 0.67$ nm.]

(5.2) The following data gives the experimental relation between stress and strain for a piece of rubber at 293 K.

Strain	Force/unstrained area (N mm^{-2})	Strain	Force/unstrained area (N mm^{-2})
0.000	0.000	4.973	1.986
0.162	0.152	5.461	2.313
0.270	0.246		
0.433	0.327	6.190	3.050
0.678	0.420	6.403	3.448
0.950	0.489	6.699	3.811
1.358	0.605	6.914	4.151
1.657	0.697	7.019	4.503
2.338	0.882	7.151	4.878
2.964	1.067	7.256	5.242
3.480	1.253	7.361	5.605
4.350	1.613	7.489	6.321

    a) Use the data at small strains to calculate the density of cross-links.
    b) Plot the data over the full range of strains, together with the prediction of rubber elasticity assuming the density of cross-links you calculated in the previous part.
    c) Discuss any discrepancies between theory and experiment.

(5.3) A certain polymer has a stress relaxation modulus given by

$$G(t, T) = \sum_{i=1}^{N} G_i \exp\left(-\frac{t}{\tau_i(T)}\right),$$

where $\tau_i(T) = A_i \exp(H/RT)$, with H constant for all relaxation times and the G_i and A_i are independent of temperature. Show that the shift factor a_T can be written

$$\ln a_T = \frac{H}{R}\left(\frac{1}{T_0} - \frac{1}{T}\right)$$

where T_0 is the reference temperature.

(5.4) At a temperature of 420 K, where a particular polymer (of density 1.06×10^3 kg m^{-3} and relative molecular mass 50 000) is rubbery, the storage modulus is found to be 2×10^5 Pa. What information concerning the architecture of the molecules does this result provide? How would you expect the modulus to change if the relative molecular mass were (a) doubled, (b) decreased by a factor of 5?

(5.5) For an entangled melt of polybutadiene, the plateau value of the shear modulus is 1.15×10^6 Pa, and the zero shear viscosity η_0 as a function of the degree of polymerisation N and the temperature (in kelvin) T may be written

$$\eta_0 = 3.68 \times 10^{-3} \exp\left(\frac{1404}{T - 128}\right) N^{3.4} \text{ Pa s}.$$

    a) Calculate the relative molecular mass between entanglements.
    b) Explain why the viscosity has this functional form.
    c) Estimate the diffusion coefficient of polybutadiene of relative molecular mass 100 000 at 298 K.

[The density of polybutadiene is 900 kg m^{-3}, the statistical step length is 0.65 nm, and the relative molecular mass of a monomer unit is 54 g mol^{-1}.]

Gelation

6.1 Introduction

A **gel** is a material composed of subunits that are able to bond with each other in such a way that one obtains a network of macroscopic dimensions, in which all the subunits are connected by bonds. If one starts out with isolated subunits, and successively adds bonds, one goes from a liquid—a **sol**—to a material with a non-zero shear modulus—a **gel**. A gel has the mechanical properties characteristic of a solid, even though it is structurally disordered and indeed may contain a high volume fraction of liquid solvent. If the gel consists of linear segments joined together at cross-link points, and the linear sections are flexible and long enough to be considered as random walks, the mechanical properties of the gel may be described by the theory of rubber elasticity described in the last chapter. Gels may also go through a glass transition; gels with a high density of short rigid segments are likely to be glassy.

Within this broad framework there are many different types of gel, with different classes of subunit, and with different types of bonding between them. For example, the subunits may be multi-functional monomers, which are connected together by covalent bonds to form a three-dimensional network. An example would be an epoxy resin. Alternatively, the subunits may themselves be linear polymers, which are connected together by covalent cross-links to form a rubber. Linear polymers may be connected by physical, rather than chemical, bonds, giving a **thermoreversible** gel such as gelatin. The subunits themselves may be colloidal aggregates of many molecules, which then are linked together by physical interactions.

Is there anything that unites these very disparate classes of materials? One common theme is that they all undergo a transition from the sol state to a gel by a process of increasing the number of bonds between subunits until there is a macroscopic network of subunits which are all connected together. This transition is known as the **gelation** transition or the **sol–gel** transition. As we shall see, it has some similarities with the thermally driven phase transitions that form such a recurring theme throughout this book. Steered by that analogy, one may ask if there are universal features of the sol–gel transition that are common throughout this wide variety of systems. We shall see that there is now reason to suppose that there is some degree of universality, though opinion is still divided as to how useful this is. Nonetheless, the sol–gel transition provides a valuable framework for us to discuss this important class of soft matter.

Thus we proceed to ask two questions, one specific, and one general.

1. What is the nature of the bonds between the subunits and how does a given system form a gel?

We will see that although there are some common themes there are a variety of detailed mechanisms for making a gel.

2. What are the general features of this kind of transition between a liquid (a sol) and a solid (a gel)?

This can be thought of as a sharp transition, with strong analogies to phase transitions driven by temperature changes, and as for the case of phase transitions we find elements of universality connecting the behaviour of apparently different systems.

6.2   Classes of gel

We can divide gels into two classes, depending on the nature of the bonds that link the subunits.

Chemical gels—the bonds linking the subunits are covalent chemical bonds.
Physical gels—the bonds linking the subunits are physical interactions.

Physical gels are sometimes also known as thermoreversible gels, as usually the physical interactions are of the form that are disrupted by increasing the temperature. One can melt a jelly, but not a cured epoxy resin.

6.2.1   Chemical gels

To make a chemical gel, one needs multi-functional units that can be linked by chemical bonds to make a three-dimensional network. The idea is illustrated in Fig. 6.1. Subunits that are difunctional cannot make a network; they can only make a collection of linear polymers. Among examples of systems which form this kind of chemical gel are:

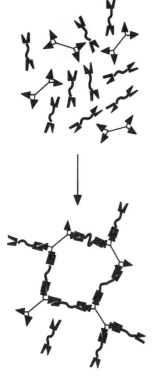

Fig. 6.1 Schematic of a thermosetting gel. The system consists of a mixture of short chains with reactive groups at each end, and cross-linker molecules, each with four functional groups capable of reacting with the ends of the chains. As the reaction proceeds the chains are linked together by the cross-linker to form an infinite network.

- **Thermosetting resins.** Materials such as epoxy resins are formed from a resin, which is a short polymer with reactive groups on both ends, and a hardener, which consists of a multi-functional molecule that can react with up to four resin end groups. When the resin and hardener are mixed the resin molecules are linked together to form a three-dimensional network. The increase in effective relative molecular mass of the growing clusters generally leads to a transition into the glassy state, so when cured epoxy resins are hard, stiff materials.

- **Sol–gel glasses.** Certain organic derivatives of silicon oxide and metal oxides are soluble in organic solvents, and in the presence of water will link together following the hydrolysis of their organic groups. In this way gels which are chemically similar to inorganic glasses may be formed by solution processing rather than by melting silica and metal oxides at high temperatures.

In both of these cases, we are forming a gel by starting out with small units which polymerise. We can also begin with long, linear polymers which we subsequently cross-link. This gives us another category of chemical gels:

- **Vulcanised rubbers.** The process on which the rubber industry was founded relied on cross-linking natural rubber—the linear polymer polyisoprene—using sulphur to yield a tough, elastic material. Modern technology uses different cross-linking agents, but the principle is the same. The basic idea is sketched in Fig. 6.2. One starts with an entangled melt of linear polymers; adjacent segments are randomly linked. As the density of cross-links increases we expect the modulus of the material to increase according to eqn 5.29. At very high cross-link densities the material becomes glassy.

6.2.2 Physical gels

In a physical gel, units are joined together by bonds which are physical in character, not chemical. These bonds can generally be broken by heating the system, so this type of gel is known as thermoreversible. Mechanisms by which such reversible cross-links can be formed include the following:

- Microcrystalline regions. In a number of polymer solutions, in regions where polymer chains meet they may begin to form small crystalline regions linking more than one chain (see Fig. 6.3). These form cross-links, but if the system is heated above the melting point of the crystallites then the cross-links will be broken, to reform again if the solution is cooled down again. The most familiar everyday example is table jelly, which is made from gelatin. Gelatin is the product of chemically degrading collagen, the group of proteins which form the major structural components of connective tissue. In its native state, the protein chains of collagen form a triple helix structure (see Fig. 6.4). When a solution of gelatin is cooled, chains come together to form little regions of triple helix; these form junctions effectively cross-linking the chains together. Because this process amounts effectively to forming small crystalline regions, there is a fairly well-defined melting point.
- Microphase separation. As we shall see in Chapter 8, a polymer in which two chemically different polymer segments are linked together covalently—a **block copolymer**—can **microphase separate**. This is illustrated in Fig. 6.5; here we envisage triblock copolymers, in which a long section of a rubbery polymer such as polybutadiene has attached at either end shorter blocks of a glassy polymer such as polystyrene. The polystyrene blocks will microphase separate into small spherical domains; because the polystyrene is glassy at room temperature the polystyrene ends will be firmly anchored into the domains, which will thus act as cross-link points for the longer, rubbery polybutadiene chains. However, if the material is heated above the glass transition temperature the polystyrene ends are no longer firmly held within the domains and the material becomes a melt. Such materials are known commercially as **thermoplastic elastomers**, and are used for applications such as shoe soles.

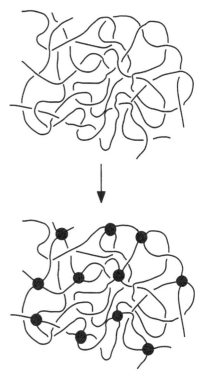

Fig. 6.2 Schematic of a vulcanisation reaction. The system consists of a mixture of long chains. Initially, the chains are entangled but not covalently linked. The reaction proceeds by chemically linking adjacent chains, leading to the formation of an infinite network.

6.3 The theory of gelation

6.3.1 The percolation model

A curious feature of gelation is the fact that as we create more bonds in a continuous way, the macroscopic properties of the gel change

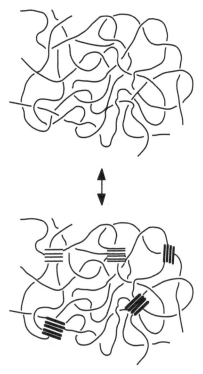

Fig. 6.3 Thermoreversible gelation by the formation of microcrystals. At low temperatures (bottom) adjacent chains form small crystalline regions which act as cross-links. Above the melting temperature, the cross-links disappear (top).

Fig. 6.4 Collagen molecules in their native conformation. Each chain is a left-handed helix; the three chains assemble to form a right-handed superhelix.

discontinuously: when a certain fraction of bonds has been made, the sample changes abruptly from a liquid- to a solid-like material. We can make a simple model that exhibits this kind of behaviour. Although the microscopic details of gelation in the many different systems that exhibit the phenomenon may be different, we might hope that there are some generic features of the transition that will be captured in a simple model. This model is the **percolation** model. It is illustrated in Fig. 6.6. We imagine an array of points which can be joined by bonds. As more bonds are added at random, clusters of points are formed, until at some stage we form a cluster that spans the entire lattice. This marks the gelation point

Among the questions we need to ask are:

- what is the fraction of bonds that need to be made in order to obtain an infinite cluster;
- how does the average cluster size vary with the fraction of bonds;
- when an infinite lattice is formed, what proportion of bonds belong to the infinite lattice as a function of fraction of bonds; and
- if the bonds represented a physical network, how would the elasticity of the network depend on the fraction of bonds?

It turns out that these questions cannot be answered by analytical methods for the percolation problem on a simple lattice as illustrated in Fig. 6.6. The problem is, however, extremely well suited to computer simulation and many results have been thus obtained. Before considering these general results, let us consider another special model which can be solved analytically.

6.3.2   The classical theory of gelation—the Flory–Stockmayer model

In this model, we consider connecting points on a **Cayley tree**. Thus we start with one point, which can connect to z other points. Each of these other points can in turn connect with z further points, and so on to infinity. This is illustrated in Fig. 6.7.

Suppose that the probability that a bond is made is f (i.e. the fraction of reacted bonds is f and we assume that each bond is independent of every other bond). Each monomer in the nth generation will be linked to $f(z-1)$ monomers in the $(n+1)$th generation. Thus if we count the number of bonds in the cluster out to the nth generation, N, we find

$$N \sim f(z-1)^n. \tag{6.1}$$

So as the number of generations n goes to infinity we find two types of behaviour, depending on whether f is greater than or less than a critical value f_c:

$$\text{if } f < f_c \quad N \to 0 \text{ as } n \to \infty \tag{6.2}$$

$$\text{if } f > f_c \quad N \to \infty \text{ as } n \to \infty \tag{6.3}$$

where f_c defines the **percolation threshold**, and is given by

$$f_c = \frac{1}{z-1}. \tag{6.4}$$

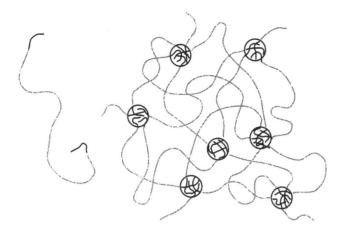

Fig. 6.5 A triblock copolymer (left) can form a thermoplastic elastomer. The end blocks microphase separate to form small, spherical domains. When these domains are glassy they act as cross-links for the rubbery centre blocks; the rubber can be returned to the melt state by heating above the glass transition of the end blocks.

Thus for $f < f_c$, below the percolation threshold, we have a solution of finite clusters—a sol. As we approach the percolation threshold, the mean size of the clusters diverges, and properties like the viscosity, which depend on the mean cluster size, also diverge.

For $f > f_c$, above the percolation threshold, we have an infinite cluster—a gel. This has a finite shear modulus.

It is important to recognise that above the percolation threshold, even though there is an infinite cluster, not all bonds are part of it. The fraction of bonds that are part of the infinite cluster is called the **gel fraction** and may be calculated in the following way.

We refer to the diagram in Fig. 6.8. Suppose that the probability that a certain site is connected to infinity by a continuous path of bonds is P, and the probability that a site is not connected to infinity by **one specified branch** is Q.

Now if we consider a neighbour to the site, we see that the probability that *none* of the neighbour's sub-branches connect to infinity is Q^{z-1}. Thus the probability that a site is connected to a neighbour, but not connected via that neighbour to infinity, is $f Q^{z-1}$.

This allows us to write a recursive relation for Q. If a site is not connected to infinity by a specified branch this is because *either* the site is not connected to the neighbour in that branch, *or* the site is connected to that neighbour, but that neighbour is not connected to infinity. Thus we can write

$$Q = 1 - f + f Q^{z-1}. \qquad (6.5)$$

Taking the simplest case of $z = 3$ this gives us a quadratic for Q with the solutions

$$Q = 1 \quad \text{or} \quad \frac{1-f}{f}. \qquad (6.6)$$

Now we can find P, the fraction of bonds connected to infinity.

Q^z is the probability that a site is not connected to infinity, and $f Q^z$ is the probability that a site is connected to a specified neighbour, but not connected to infinity. This is equal to $f - P$, the fraction of bonds that are reacted, but which do not form part of the infinite cluster.

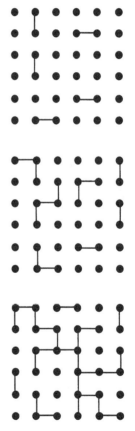

Fig. 6.6 The percolation model. We start with array of points, to which bonds are added at random (top). As more bonds are added, clusters of points are formed (middle), which ultimately join to form a cluster which spans the entire system (bottom).

Thus $f - P = fQ^z$. Once again taking the special case of $z = 3$ and using eqn 6.5 we find

$$\frac{P}{f} = 1 - \left(\frac{1-f}{f}\right)^3 \quad \text{for } f > f_c \tag{6.7}$$

or

$$\frac{P}{f} = 0 \quad \text{for } f < f_c. \tag{6.8}$$

The quantity P/f is the fraction of reacted bonds that form part of the infinite network–the **gel fraction**.

We plot this function in Fig. 6.9. The gel fraction abruptly rises from zero at the gel point; in fact we can see that close to the gel point the gel fraction can be expanded as a power law function of the distance away from the gel point. To leading order this gives

$$\frac{P}{f} = 3(f - f_c) + O(f - f_c)^2. \tag{6.9}$$

This kind of power law divergence of quantities close to a critical point is characteristic of phase transitions, emphasising the relationship between gelation and a thermal phase transition.

6.3.3   Non-classical exponents in the percolation model

In the study of thermal phase transitions, one generally finds that one can use mean field models to predict phase transitions, and that these models predict that certain quantities diverge near the transition according to a power law, but that the predicted exponents in the power law are incorrect (Chaikin and Lubensky 1995). These shortcomings are due to the neglect of fluctuations in mean field theories.

The same thing is true for percolation. The mean field model—in this case the classical Flory–Stockmayer theory described above—predicts the existence of the critical point, and predicts power law divergences near the critical point. For example, the gel fraction can be written $P/f \sim (f - f_c)^\beta$ where the exponent $\beta = 1$ in the Flory–Stockmayer theory. Monte Carlo studies of bond percolation in three dimensions give, by contrast, a value of $\beta = 0.41$. The classical theory is incorrect in detail because it neglects the possibility that **closed loops** can be formed. In general, one can expect that close to the gelation point there will be similar discrepancies between any predictions of the classical theory and the results of experiment or computer simulation.

6.3.4   The elasticity of gels

If we reduce the number of bonds towards the critical fraction f_c, we would expect the modulus of the gel (either the shear modulus or the Young modulus)

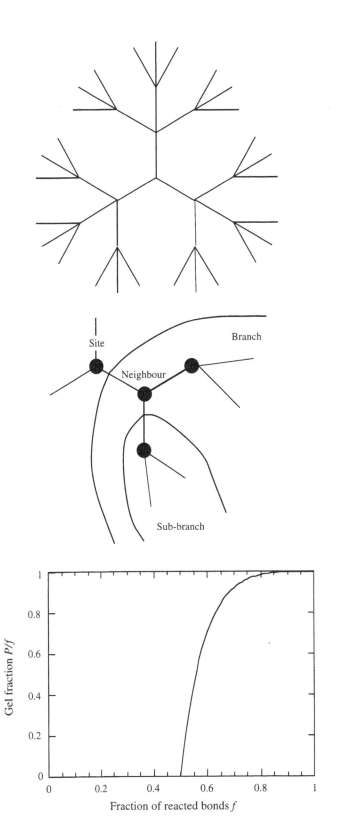

Fig. 6.7 Three generations of a Cayley tree.

Fig. 6.8 Definitions of branches and neighbours on a Cayley tree.

Fig. 6.9 The gel fraction in the classical model of gelation for a coordination number $z = 3$, as given by eqns 6.7 and 6.8.

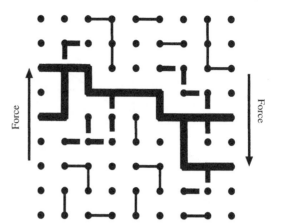

Fig. 6.10 The effect of dangling ends on the shear modulus of a gel near the gel point. The bonds shown with dashed lines are part of the infinite network, but do not contribute to the elastic modulus.

to go to zero. At first, one might imagine that it would go to zero in the same way as the gel fraction. However, further thought suggests that this cannot be the case, because of the effect of **dangling ends**. This is illustrated in Fig. 6.10; here the bonds that are shown with dashed lines are part of the infinite network, but because they are only connected to the network by one end they cannot contribute to the transfer of stress across the sample.

There is a possible analogy here with the conduction of electricity through a random network. Here one imagines making a circuit by connecting randomly selected adjacent points on a lattice with resistors. Resistors that are not in a continuous pathway from one contact to the other do not contribute to the conductance of the network Σ. Both numerical studies and experiments show that the conductance goes to zero at the gelation point with a power law with an exponent μ:

$$\Sigma \sim (f - f_{c})^{\mu}. \tag{6.10}$$

For the 3d bond percolation model the exponent $\mu \approx 2.0$, while for the Flory-type models one finds the larger value of $\mu = 3$.

Experiments on the modulus of networks just above the gel point do seem to show power law behaviour, though the exponent is not always in agreement with that predicted for the conductance of random networks. It is possible that the correspondence between the modulus of a network and the conductance of a resistor network is not exact, because the bending rigidity of the bonds must be important in real gelling systems.

Further reading

Percolation theory and its applications are described in Stauffer and Aharony (1994). Both the classical theory of polymer gelation and the application of percolation theory to the polymer problem are described in de Gennes (1979).

Exercises

(6.1) In a certain chemical cross-linking reaction involving a monomer that can react at three sites, the degree of reaction f obeys the second-order rate law

$$\frac{\mathrm{d}f}{\mathrm{d}t} = k(1 - f)^2,$$

where the rate constant k has the value $4 \times 10^{-4}\,\mathrm{s}^{-1}$. Use the Flory–Stockmayer theory to calculate

a) the time at which the gel point is reached,

b) the time after which three-quarters of the monomers have been polymerised,

c) the time after which three-quarters of the monomers form part of the infinite network.

(6.2) In an experiment to test the application of the theory of percolation to gelation, the gel fraction is determined when the fractional extent of reaction is a small degree Δf larger than its value at the gel point.

a) Is the value of the gel fraction at a fractional extent of reaction $\Delta f/2$ larger or smaller when predicted by percolation theory than the value predicted by Flory–Stockmayer theory?

b) By what factor do the two predictions differ?

7 Molecular order in soft condensed matter—liquid crystallinity

7.1 Introduction

Most of the materials that are studied in the field of 'hard' condensed matter physics—metals, semiconductors, and ceramics—are crystalline; the atoms or molecules of which they are composed are arranged with near-perfect long-ranged order over distances that are many orders of magnitude greater than the distance between molecules. Single crystals of metals or semiconductors of macroscopic size are not uncommon, but even where these materials are described as 'polycrystalline' the fraction of the molecules that does not partake in the long-ranged order is very small (though, of course, this small fraction of atoms associated with grain boundaries and defects such as dislocations may have an effect on bulk properties out of proportion to their number).

The situation in soft condensed matter is rather different. Crystallinity—involving full long-ranged positional order—is important in soft matter, but in most soft matter systems the degree of molecular ordering falls somewhere between the full positional order of a single crystal and the complete positional disorder of a liquid or a glass. In fact, there are two distinctly different types of intermediate order in soft matter systems:

1. **Liquid crystallinity**. These are **equilibrium** phases in which molecules are arranged with a degree of order intermediate between the complete disorder of a liquid and the long-ranged, three-dimensional order of a crystal.
2. **Partial crystallinity**. This is a **non-equilibrium** state of matter in which the system is prevented from reaching its equilibrium state of full long-ranged order for kinetic or other reasons, and in which microscopic regions of crystalline order coexist with disordered regions, often in a complex hierarchical structure.

Liquid crystalline phases are found in

- certain organic compounds with highly anisotropic molecular shapes—these are the materials used in liquid crystalline displays;
- polymers composed of units having a high degree of rigidity, either in the backbone or attached to the backbone as side chains;
- polymers or molecular aggregates which form rigid rod-like structures in solution.

Partial crystallinity is typical of many flexible polymers, such as polyethylene or poly(ethylene terephthalate).

In this chapter we discuss liquid crystallinity, and move on to consider partial crystallinity in polymers in Chapter 8.

7.2 Introduction to liquid crystal phases

A crystal has long-ranged, three-dimensional, positional order, while a liquid has neither positional order nor orientational order. Liquid crystalline phases possess order intermediate between these two extremes. The most disordered type of liquid crystalline phase is the **nematic** phase, which has no positional order, but in which the molecules are, on average, oriented about a particular direction, called the **director**. The transition between the isotropic phase and the nematic phase is sketched in Fig. 7.1. The absence of positional order in a nematic phase means that if one plotted the centres of mass of the molecules the arrangement would be indistinguishable from an isotropic liquid; the only ordering is in the orientation of the molecules, and even this ordering is, as sketched, not perfect. This point will emerge more clearly when we consider statistical mechanical theories of the transition from an isotropic to a nematic liquid, in the next section.

A variant of the nematic phase occurs in systems where the system is composed of molecules which are **chiral**; that is, in which the molecule differs from its own mirror image. In these systems there may be a slight tendency for neighbouring molecules to align at a slight angle to one another. This weak tendency leads the director to form a helix in space, with a well-defined pitch which is much longer than the size of a single molecule. These phases are called **chiral nematics**, or perhaps more commonly **cholesterics**. In many cases the pitch of the helix is of the same order as the wavelength of light, and so these materials can display striking optical effects.

There are still more phases that are intermediate in order between nematics and crystals. In a **smectic** phase, the molecules arrange themselves in sheets. Within each layer, the molecules are aligned, but have no positional order. Thus in going from a nematic to a smectic phase, we go from a situation of no positional order to long-range positional order in one dimension only (Fig. 7.2). Two common subclasses of smectic ordering are the **smectic A** phase, in which the director is parallel to the layer normal, and the **smectic C** phase, in which

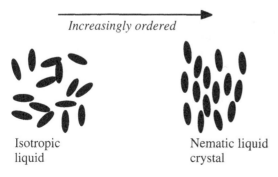

Increasingly ordered

Isotropic
liquid

Nematic liquid
crystal

Fig. 7.1 A sketch of the transition between the isotropic liquid phase, in which there is neither positional nor orientational order, and a nematic phase, in which there is orientational order, but still no positional order.

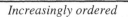

Increasingly ordered

Nematic liquid
crystal

Smectic A liquid
crystal

Fig. 7.2 A sketch of the transition between the nematic phase, in which there is orientational order but no positional order, and a smectic A phase, in which there are orientational order and long-ranged positional order in one dimension.

Fig. 7.3 A sketch of a columnar phase, in which disk-shaped molecules are arranged with orientational order and long-ranged positional order in two dimensions. Within each column there is no long-ranged positional order.

Table 7.1 Degrees of order in liquid crystalline phases.

Phase	Positional order	Orientational order
Liquid	None	None
Nematic	None	Yes
Smectic	One-dimensional	Yes
Columnar	Two-dimensional	Yes
Crystalline	Three-dimensional	Yes

the director and the layer normal make an angle. Thus a smectic C phase is made up of layers of tilted molecules.

Finally, it is possible to have a phase which has positional order in two dimensions as well as orientational order. This kind of phase is found in molecules that are disk-like, rather than rod-like; in a **columnar** (or **discotic**) phase such molecules stack into long columns. There are a number of different columnar phases in which there are different degrees of long-ranged order in the arrangement of the columns. One such phase is illustrated in Fig. 7.3. Within each column there is no long-ranged order in the position of the molecules, but the columns arrange themselves into a regular two-dimensional hexagonal lattice.

These different levels of positional and orientational order are summarised in Table 7.1.

7.3 The nematic/isotropic transition

The simplest and least ordered liquid crystal phase is the nematic phase, in which there is no positional order, but in which there is long-ranged order of the direction of the molecules. In going from an isotropic state, in which both position and orientation are random, to a nematic state, in which position is random but there is a preferred orientation, there must be a reduction in the orientational entropy of the system. So in order for the nematic state to have a lower free energy than the isotropic state, there must be another term in the free energy which favours orientation. Then, as the temperature changes, the relative importance of the two terms changes, leading to a phase transition.

How can we describe the state of orientational order of a molecule in a quantitative way? For a rod-like molecule we can introduce a single preferred direction, the director, and we introduce an **orientation function** $f(\theta)$; $f(\theta) \, d\Omega$ is the fraction of molecules in a solid angle $d\Omega$ which are oriented at an angle of θ to the director. For a completely randomly oriented molecule, there is an equal chance that the molecule points anywhere in a solid angle of 4π, and $f(\theta)$ is constant. For a more ordered system the function becomes more peaked around the angles 0 and π, as shown in Fig. 7.4. In all known nematics, the directions 0 and $+\pi$ are identical, so $f(\theta) = f(\pi - \theta)$.

The distribution function contains all the information about the state of order in the material, but it would be convenient to represent this state of order not as a function but as a simple number—an **order parameter**—which took the value 0 for complete disorder, and 1 for complete order. One might think of taking the average of $\cos \theta$, but this is zero because $f(\theta) = f(\pi - \theta)$. Instead we must take another average: $\frac{1}{2}\langle 3 \cos^2 \theta - 1 \rangle$ has the right properties. Thus we define the order parameter S by

$$S = \frac{1}{2}\langle 3 \cos^2 \theta - 1 \rangle = \int \frac{1}{2}(3 \cos^2 \theta - 1) f(\theta) \, d\Omega. \tag{7.1}$$

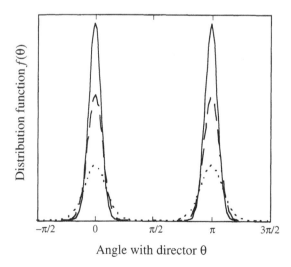

Fig. 7.4 The distribution function $f(\theta)$ for a nematic phase with various degrees of order. The order parameter S takes the value 0.82, 0.71, and 0.44 for the solid, dashed, and dotted lines respectively.

Why should a liquid adopt a nematic phase? In going from an isotropic state to a state of orientational order, there must be a loss of the entropy associated with the freedom of a molecule to be oriented in any arbitrary direction. If the nematic phase is to be at equilibrium, the positive contribution to the free energy arising from this loss of orientational entropy must be outweighed by some other factor that causes the free energy to be lowered when the molecules are aligned. This is likely to occur in melts of rod-like objects for two reasons:

(1) favourable attractive interactions arising from van der Waals forces between the molecules will be maximised when they are aligned;
(2) it is easier to pack rod-like molecules when they are aligned.

The first factor is perhaps most important for melts of relatively small molecules which form nematic phases; the second factor is the major factor underlying the transitions that occur as a function of concentration for very long rigid molecules and supramolecular assemblies. In both cases, simple statistical mechanical theories can be formulated on the basis of these ideas. These theories, which yield predictions about the nature of the transition between the isotropic and nematic states, are both mean field theories, and as such are similar in spirit to theories introduced elsewhere in this book to describe other phase transitions.

The starting point for both theories is to write down an expression for the entropy lost when molecules become oriented. We can write the contribution to the entropy of a molecule due to its orientational freedom using the Boltzmann formula as

$$S_{\text{orient}} = -k_{\text{B}} \int f(\theta) \ln f(\theta) \, d\Omega. \tag{7.2}$$

In the isotropic state, $f(\theta) = 1/4\pi$, so the change in entropy per molecule on going from the isotropic state to an ordered state is given by

$$\Delta S = -k_{\text{B}} \int f(\theta) \ln[4\pi f(\theta)] \, d\Omega. \tag{7.3}$$

In the first theory we consider, which is known as the **Maier–Saupe theory**, we make the phenomenological assumption that the energetic interaction between molecules is simply a quadratic function of the order parameter, so we write the total free energy change per molecule on going from the isotropic to the nematic state as

$$\Delta F = -uS^2/2 + k_{\text{B}}T \int f(\theta) \ln[4\pi f(\theta)] \, d\Omega, \tag{7.4}$$

where u is a parameter that expresses the strength of the favourable interaction between two neighbouring molecules. Of course, S is defined in terms of the distribution function $f(\theta)$. What we now need to do is find the function $f(\theta)$ which minimises the free energy.

We do this in two stages. Firstly, for a given value of the order parameter S, we find the most probable distribution function $f(\theta)$ by maximising the entropy associated with $f(\theta)$ subject to the constraint of a fixed value of S. From this most probable distribution function we can calculate the entropy. In this way we can find the orientational entropy as a function of the order parameter.

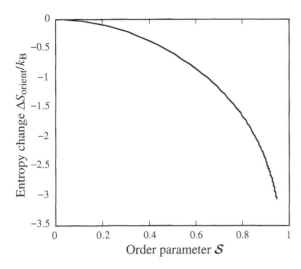

Fig. 7.5 The change in orientational entropy per molecule on going from an isotropic state to an ordered state as a function of the order parameter S.

Thus we need to find the function $f(\theta)$ that gives a stationary value of the integral $\int f(\theta) \ln f(\theta) \sin \theta \, d\theta$ subject to the constraint that the integral $\int \frac{1}{2}(3 \cos^2 \theta - 1) f(\theta) \sin \theta \, d\theta = S$ is a constant. The Euler–Lagrange equation resulting from this problem in the calculus of variations is

$$\ln f + \frac{3\lambda}{2} \cos^2 \theta + 1 - \frac{\lambda}{2} = 0, \tag{7.5}$$

which has the solution

$$f(\theta) = \exp(3 \lambda \cos \theta^2), \tag{7.6}$$

where λ is the Lagrange multiplier that sets the value of the order parameter S. Now, for a given value of λ, and thus a given value of S, we can evaluate the entropy using eqn 7.3. The resulting curve showing the change in orientational entropy per molecule ΔS_{orient} on going to an oriented state with order parameter S is shown in Fig. 7.5.

We can now plot the free energy as predicted by eqn 7.4 as a function of order parameter for various values of $u/k_B T$. This is shown in Fig. 7.6. For relatively small values of $u/k_B T$, the minimum free energy is found for a value of the order parameter of zero; here the free energy is dominated by the orientational entropy term, and the equilibrium state is isotropic. But as the coupling parameter is increased a minimum of the free energy is found for a non-zero value of S: the equilibrium phase is nematic. The critical value of $u/k_B T$ for the transition is around 4.55.

By calculating the value of the order parameter S as a function of $u/k_B T$ we can investigate the character of the transition. This is shown in Fig. 7.7; at a value of $u/k_B T = 4.55$ there is a discontinuous change of the order parameter from $S = 0$ to $S = 0.44$. This is the nematic–isotropic phase transition; because it is a discontinuous change it is a **first-order** phase transition. However, as the minimum in free energy at the transition is rather shallow, fluctuations at the transition will be important. The transition should therefore be considered to be only weakly first order, and the change in degree of order at the transition is usually not very great.

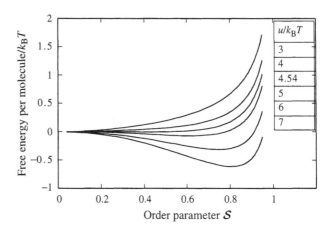

Fig. 7.6 The free energy as a function of the order parameter S for various values of the coupling parameter $u/k_B T$, as given by the Maier–Saupe theory (eqn 7.4).

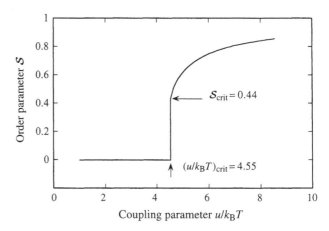

Fig. 7.7 The order parameter S as a function of the coupling parameter $u/k_B T$, as given by the Maier–Saupe theory. There is a weak first-order phase transition at $u/k_B T = 4.55$.

In order to compare the predictions of the Maier–Saupe theory with experiment we would need to have some theory about the way in which the coupling parameter u varied with temperature. The simplest assumption is that u is independent of temperature; this would be the case if the coupling arose entirely from van der Waals forces. This turns out to be quite a reasonable first approximation for small-molecule liquid crystals. Figure 7.8 compares experimentally measured ordered parameters for the molecule p-azoxyanisole (PAA) with the prediction of Maier–Saupe theory assuming that u takes the temperature-independent value that reproduces the experimentally observed transition temperature (i.e. this is a one-parameter fit). There is quite good qualitative agreement. While the theory captures the relatively small degree of order at the transition and gives a good account of the development of order with decreasing temperature, there are clearly systematic deviations from the predictions of theory, particularly close to the transition.

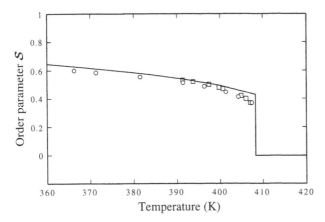

Fig. 7.8 The order parameter S as a function of temperature for p-azoxyanisole (PAA), as measured by refractive indices (circles, from S. Chandrasekhar and N.V. Madhusudana, *Appl. Spectrosc. Rev.*, **6**, 189 (1972)) and diamagnetic anisotropy (squares, from H. Gasparoux, B. Regaya, and J. Prost, *C.R. Acad. Sci.*, **272B**, 1168 (1971)). The solid line is the prediction of Maier–Saupe theory assuming u is independent of temperature and takes a value which reproduces the experimental transition temperature.

There are a number of reasons why there are discrepancies between the experimental data and the predictions of Maier–Saupe theory. Two such possible factors are:

1. Intrinsic temperature dependence of u. This could arise, for example, because the excluded volume interaction is significant.
2. Neglect of fluctuations. The Maier–Saupe theory is a mean field theory, and like all such theories it neglects the effects of fluctuations in the order parameter. These are likely to become important close to the transition point.

7.4 Distortions and topological defects in liquid crystals

7.4.1 Generalised rigidity and the elastic constants of a nematic liquid crystal

Why is it that when one pushes the end of a lever or a beam, the force applied is transmitted from this end to the other? We are so used to the idea that solids are rigid that we forget that the property of rigidity is rather mysterious. After all, we now know that a solid beam is mostly empty space, with atoms or molecules locked into their positions by a subtle balance of forces. It is not just the strength of the interatomic interactions that allows macroscopic forces to be transmitted over macroscopic distances. After all, the density of a solid is usually very similar to the density of its melt, and the total interaction energies in the two situations are very similar, yet the difference in behaviour is qualitative: solids are rigid, and liquids are not. The difference is in the long-ranged order of the solid; if one moves the position of an atom at one end of a rod, an atom at the other end somehow knows it has to try and move in order to maintain the

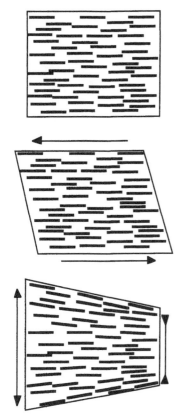

Fig. 7.9 Deformations of a nematic liquid crystal. The shear deformation (middle) does not perturb the long-ranged orientational order, and thus is not opposed by any increase in elastic energy. The splay deformation (bottom) does perturb the long-ranged order, and is opposed by an increase in elastic energy.

absolute precision of the crystalline order. Thus whenever one has long-ranged order of any kind, one has some kind of rigidity which ensures that the system does its best to maintain its long-ranged order when a part of it is perturbed.

In a solid, this rigidity is described by the theory of elasticity. If we apply a stress to a crystal it will deform in response, but this deformation leads to a perturbation of its long-ranged order. This deformation leads to an increase in the energy of the solid proportional to the square of the size of the deformation, as long as the deformation is relatively small (Hooke's law), and if we release the stress the crystal will relax back to its original shape. This elastic behaviour is in contrast to the behaviour of a liquid, which shows no rigidity; if we apply a stress the liquid flows.

In a liquid crystal if an applied stress leads to a deformation that perturbs any long-ranged order that the system possesses, then the deformation will be opposed by an increase in elastic energy. On the other hand, if the applied stress leads to a deformation that does not perturb the long-ranged order, then there will be no increase in elastic energy and the material will respond by flow. Thus the essence of the mechanical response of a liquid crystal is that it has an elastic response to some types of deformation, and a liquid-like response to others. This is illustrated for a nematic liquid crystal in Fig. 7.9. This shows that a nematic liquid crystal will flow like a liquid in response to a simple shear stress, because the resulting deformation leaves the long-ranged orientational order unchanged. On the other hand, more complicated types of deformation, such as the splay deformation illustrated, do lead to an increase in elastic energy. We can define elastic constants for this type of deformation.

In the continuum limit, we can characterise a nematic liquid crystal in a state of deformation by the vector field $\mathbf{n}(\mathbf{r})$ giving the director at every point $\mathbf{r}$. In analogy to Hooke's law, we expect the elastic energy to be proportional to terms in the square of space derivatives of $\mathbf{n}(\mathbf{r})$. Formally, we can enumerate the space derivatives of $\mathbf{n}(\mathbf{r})$, which form a second-rank tensor; by exploiting symmetry properties one can show that there are only three independent elastic constants (de Gennes and Prost 1993). We shall merely state the result, and then discuss the physical significance of the terms.

The elastic energy of distortion per unit volume of a nematic liquid crystal, F_d, can be written

$$F_d = \frac{1}{2}K_1(\nabla \cdot \mathbf{n})^2 + \frac{1}{2}K_2(\mathbf{n} \cdot \nabla \times \mathbf{n})^2 + \frac{1}{2}K_3((\mathbf{n} \cdot \nabla) \cdot \mathbf{n})^2, \quad (7.7)$$

where K_1, K_2, and K_3 are the three elastic constants. These three constants correspond to three fundamental types of deformation in nematic liquid crystals: **splay**, **twist**, and **bend**. These are illustrated in Fig. 7.10.

In practice, we find that values for the elastic constants in typical small-molecule liquid crystals are of order 10^{-12} N. All three constants are of the same order of magnitude, though the bending constant K_3 is generally somewhat larger than the other two.

7.4.2   Boundary effects

In discussing the elastic constants of nematic liquid crystals, we glossed over the important practical question of how it is that one can impose a distortion on a nematic liquid crystal. It turns out that an easy and practically important method is to exploit the property that many surfaces have of imposing a preferential state of ordering on the nematic liquid crystal.

In fact, virtually any surface will impose some sort of ordering on the director of an adjacent liquid crystal. There are two important cases:

1. **Homeotropic**—alignment perpendicular to the surface. This can be achieved by arranging for the surface to be coated by a surfactant molecule.
2. **Homogeneous**—alignment parallel to the surface. This is the more usual case. For a free surface any direction in the plane of the surface may be allowed, while for a solid substrate a particular direction in the plane may be imposed by the crystalline structure of the surface. A direction may also be imposed simply by rubbing a glass or plastic surface. The mechanism by which this rubbing process works is still a little obscure. In the case of the most technologically important process, in which polymers of the polyimide family are rubbed with a velvet cloth, it seems that the rubbing causes an alignment of the polymer chains. The molecules of the liquid crystal then tend to be aligned with the chain direction.

The important general point to remember about these effects is that the sensitivity to alignment at the boundary is a consequence of the **broken symmetry** of the nematic phase. In the bulk, the energy of a nematic monodomain is independent of the orientation of the director. Thus the smallest perturbation that can change the relative energies of the different orientations will be sufficient to impose this preference on the material.

7.4.3 Disclinations, dislocations, and other topological defects

Up to this point, we have assumed that the director in a nematic phase always varies through space in a smooth way. In fact, experimental samples of liquid crystals will often contain points or lines where the orientation of the sample changes discontinuously. These topological defects are known as **disclinations**.

Two important types of disclination are illustrated in Fig. 7.11. Here the local directions of the director are drawn in the plane perpendicular to the disclination line. One can define the **strength** of the disclination by a construction illustrated in this diagram. If one draws a loop round the defect core, and one goes round the loop drawing an arrow in the direction of the local director, in disclination (a) the arrow rotates π radians in the same sense as the direction of traverse of the loop. This disclination has strength $s = +\frac{1}{2}$. In disclination (b) the rotation is also π radians, but the sense of rotation is opposite to that of the traverse of the loop; the disclination has strength $s = -\frac{1}{2}$.

Disclinations should not be present in a macroscopic sample of a liquid crystal at equilibrium, because they lead to an increase in the free energy proportional to the length of disclination line present. The origin of this energy is two-fold: there is a certain elastic energy associated with the long-ranged distortion of the director field, while right at the core of the disclination there is an increase in energy associated with the complete loss of orientational order there. Accurate calculation of the disclination energy is complex, particularly

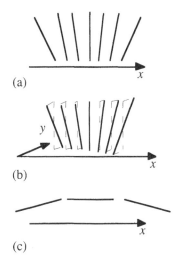

Fig. 7.10 The three fundamental deformations of a nematic liquid crystal. (a) Splay, $(\nabla \cdot \mathbf{n}) \neq 0$. (b) Twist, $\nabla \times \mathbf{n}$ parallel to $\mathbf{n}$. (c) Bend, $\nabla \times \mathbf{n}$ perpendicular to $\mathbf{n}$.

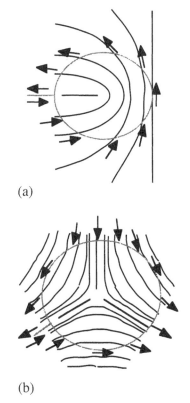

Fig. 7.11 Disclinations in a nematic liquid crystal. The disclination in (a) has strength $s = +\frac{1}{2}$, the disclination in (b) has strength $s = -\frac{1}{2}$.

if one goes beyond an approximation in which the three elastic constants, K_1, K_2, and K_3, are treated as equal. In this approximation, if we can write $K = K_1 = K_2 = K_3$, then we can show that the energy E of length L of disclination is given by

$$\frac{E}{L} = \pi K s^2 l, \tag{7.8}$$

where l is a factor which varies only logarithmically with quantities such as the size of the disclination core and the average separation of disclinations. An energy per unit length corresponds to a **line tension**; thus we can consider a disclination in some respects as behaving like a string under tension.

Disclinations correspond to defects in the state of orientational order. If we have a liquid crystal that has some degree of translational order, such as a smectic one, then we can expect also defects in translational order. Such defects, which are familiar from crystalline solids, are known as **dislocations**.

7.5   The electrical and magnetic properties of liquid crystals

The anisotropic nature of liquid crystal phases manifests itself in particularly striking ways in regard to their interaction with electromagnetic fields. This is the basis not only of the very striking optical effects that are characteristic of liquid crystal phases, but also of their crucial role in display technologies. Our discussion will mostly be confined to effects in nematics, which are by far the most important systems for practical applications.

Most of the molecules that form nematic phases have a permanent dipole, but it is found that in the absence of an applied field there is an equal probability that the dipole points in either direction—nematics are never **ferroelectric**. When a nematic phase is put in an electric field $\mathbf{E}$ the field induces a polarisation $\mathbf{P}$ which is given, in an isotropic system, by $\mathbf{P} = (\epsilon - 1)\epsilon_0\mathbf{E}$, where ϵ is the dielectric constant. In a nematic liquid crystal, the degree to which the material can be polarised depends on whether the field is applied parallel or perpendicular to the director.

This is illustrated in Fig. 7.12, which sketches the response to an applied field of a nematic liquid crystal in which the molecules have a dipole parallel to their long axis. Much more polarisation can be induced when the applied field is in the same direction as the director. The consequence is that the dielectric coefficient parallel to the director $\epsilon_\parallel$ is larger than the dielectric coefficient perpendicular to the director $\epsilon_\perp$. The situation is reversed if the dipole moment lies perpendicular to the long axis of the molecule; then we find $\epsilon_\parallel < \epsilon_\perp$.

The consequence of this is that electric fields are able to align nematic liquid crystals rather efficiently. In the case of a parallel dipole, if the director adjusts to be parallel to the applied field, the electric field within the material is reduced by the polarisation effect and as a result the total energy of the system is reduced. For a field applied in some general direction, we can write the energy in terms of the displacement $\mathbf{D}$. This is given by

$$\mathbf{D} = \epsilon_\perp\mathbf{E} + (\epsilon_\parallel - \epsilon_\perp)(\mathbf{n} \cdot \mathbf{E})\mathbf{n}, \tag{7.9}$$

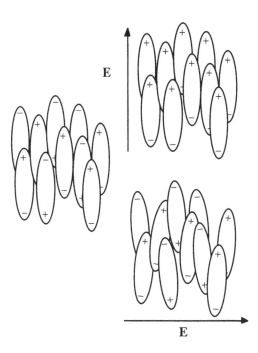

Fig. 7.12 Dielectric anisotropy in a nematic liquid crystal, in which there is a permanent electric dipole parallel to the long axis of the molecule. In the absence of field (left) there is no net dipole moment. If the field is applied parallel to the director (top right), it is relatively easy to acquire a large induced dipole, while if the field is perpendicular to the director (bottom right) the degree of polarisation is smaller. Thus the dielectric coefficient parallel to the director $\epsilon_\parallel$ is larger than the dielectric coefficient perpendicular to the director $\epsilon_\perp$.

so we can write the energy per unit volume associated with the field, F_{el}, as

$$F_{el} = -\frac{1}{4\pi} \int \mathbf{D} \cdot d\mathbf{E} = -\frac{\epsilon_\perp}{8\pi} E^2 - \frac{\epsilon_a}{8\pi} (\mathbf{n} \cdot \mathbf{E})^2, \qquad (7.10)$$

where $\epsilon_a = \epsilon_\parallel - \epsilon_\perp$ and the director is along the unit vector $\mathbf{n}$. Thus for dipoles parallel to the long axis ϵ_a is positive and the energy is reduced when the nematic liquid crystal aligns along the field.

The response of a nematic liquid crystal to a magnetic field is similarly anisotropic. Most liquid-crystal-forming molecules are **diamagnetic**; an applied magnetic field H causes a molecular current which produces a magnetic field opposing the applied field. The coefficient relating the induced magnetisation M to the field H is the susceptibility $\chi = M/H$. The susceptibility takes different values according to whether the field is applied parallel to the director ($\chi_\parallel$) or perpendicular to the director ($\chi_\perp$). Usually the difference between the two susceptibilities, $\chi_a = \chi_\parallel - \chi_\perp$, is **positive**; the director tends to line up with the direction of the magnetic field. Entirely analogously to the electrical case, we can write down the energy per unit volume associated with the magnetic field, F_{mag}, as

$$F_{mag} = -\int \mathbf{M} \cdot d\mathbf{H} = -\frac{1}{2}\chi_\parallel H^2 - \frac{1}{2}\chi_a (\mathbf{n} \cdot \mathbf{H})^2, \qquad (7.11)$$

where the director is along the unit vector $\mathbf{n}$.

7.6 The Frederiks transition and liquid crystal displays

We have seen that the alignment of liquid crystals is highly sensitive to external fields; these fields include electric and magnetic fields in the bulk, and the effects of surfaces in imposing alignment. There is a class of interesting transition effects that occur when two antagonistic aligning influences compete—these are known as **Frederiks transitions**.

Perhaps the simplest of these transitions occurs when one has a thin film of a nematic liquid crystal sandwiched between two plates. If these plates impose a strong parallel alignment effect on the liquid crystals then the liquid crystal will form a single domain. Let us now impose a field (either electric or magnetic) perpendicular to the plates that has a tendency to align the director in that direction. The surface constraint forces the director to lie parallel to the plates at the edge of the cell, while the field tends to impose a perpendicular orientation towards the centre. The only way both can be satisfied is if there is a distortion of the director field that will cost a splay energy. It is this splay energy that means that the cell will only change from one configuration to the other when a certain critical field has been applied.

To see this we can carry out a simple linear stability analysis. Let us suppose that the unperturbed director is $\mathbf{n}_0$, and that towards the centre of the cell the director is slightly perturbed and lies in the direction $\mathbf{n}_0 + \delta\mathbf{n}(z)$, where $\delta\mathbf{n}(z)$ is a small vector perpendicular to the plates. We can now write the total free energy per unit volume in terms of an elastic contribution and a contribution to the field. This has the form

$$
\begin{aligned}
F_{\text{total}} &= F_{\text{elastic}} + F_{\text{field}} \\
&= \frac{1}{2} K_1 \left(\frac{\partial \delta\mathbf{n}(z)}{\partial z} \right)^2 - \frac{\epsilon_a E^2}{8\pi} \delta\mathbf{n}(z)^2,
\end{aligned} \tag{7.12}
$$

where K_1 is the splay elastic modulus and ϵ_a is the difference in dielectric coefficients perpendicular and parallel to the field. At the boundaries $z = 0$ and $z = d$ (where d is the thickness of the cell), $\delta\mathbf{n} = 0$ because of the alignment properties of the surface, which we assume to be strong. So consider a distortion of the form

$$
\delta\mathbf{n}(z) = \delta n \sin\left(\frac{\pi z}{d} \right). \tag{7.13}
$$

If we substitute this into eqn 7.12 and integrate from $z = 0$ to $z = d$ we get the following expression for the energy:

$$
\int_0^d F_{\text{total}} \, dz = \delta n^2 \left[\left(\frac{K_1 \pi^2}{4d} \right) - \left(\frac{\epsilon_a E^2 d}{16\pi} \right) \right]. \tag{7.14}
$$

So any small distortion of the director will lead to a lowering of the free energy if the field exceeds a critical value E_{crit}, which is given by

$$
E_{\text{crit}} = \frac{2\pi}{d} \sqrt{\frac{\pi K_1}{\epsilon_a}}; \tag{7.15}
$$

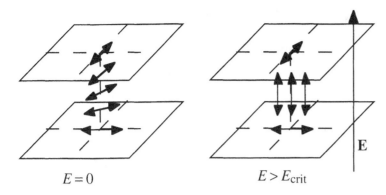

$E = 0$ $E > E_{\text{crit}}$

Fig. 7.13 The twisted nematic display.

for values of applied field less than E_{crit} the director remains parallel to the cell walls right through the cell, while for values greater than E_{crit} the director distorts in the centre of the cell to align with the field.

Very similar calculations can be made for the case of a liquid crystal in a magnetic field. In fact, observations of magnetic Frederiks transitions provide the best way experimentally to determine the elastic constants for twist, splay, and bend.

Frederiks transitions, in slightly more complicated geometries, form the basis for **liquid crystal displays**. For example, in a **twisted nematic display**, a nematic liquid crystal is confined between two plates, typically around $10\,\mu$m apart. The plates are treated to impose a parallel alignment on the liquid crystal, but the alignment directions of the two plates are perpendicular to each other. This forces the director to twist round by $90°$ between the top and bottom plate. Now if a field is applied perpendicular to the plates, above the Frederiks transition the director towards the centre of the plates will swing into alignment with the applied field (Fig. 7.13).

The critical field for the Frederiks transition in the twisted nematic geometry can be calculated by a linear stability argument similar to the one given above; the situation is slightly more complicated because bend and twist distortions are involved as well as splay distortions. The result is that the critical field is given by

$$E_{\text{crit}} = \frac{2\pi}{d}\sqrt{\frac{\pi}{\epsilon_{\text{a}}}}\left[K_1 + \frac{1}{4}(K_3 - 2K_2)\right]^{1/2}. \qquad (7.16)$$

To use the twisted nematic cell as a display device, the cell is sandwiched between crossed polarisers. In the voltage-off state, the polarisation state of incoming light is twisted round through $90°$ as it follows the changing orientation of the director through the cell. When a field substantially greater than the critical field is applied, the director through most of the cell is perpendicular to the plates and has no effect on the polarisation state of the light; the polarised light is blocked by the polariser as it leaves the cell, which appears dark.

7.7   Polymer liquid crystals

7.7.1   Rigid polymers

Polymers form an important subclass of liquid-crystal-forming materials. Polymer liquid crystal phases are important in the processing of advanced high-modulus engineering materials, like Kevlar, and they may also occur in nature in solutions of some biopolymers.

We can distinguish between two classes of liquid crystal polymer:

thermotropic liquid crystal polymers, in which the transition from an isotropic to a liquid crystalline phase is driven by changes in temperature;
lyotropic liquid crystal polymers, in which a liquid crystal phase is formed in solution, with a transition from isotropic to liquid crystalline phase driven by a change in concentration.

A polymer will have a propensity to form a liquid crystalline phase if its backbone is relatively **rigid**; liquid crystal phases are also possible for polymers with a flexible backbone with rigid units attached, but these **side-chain liquid crystals** are rather different. Such rigidity in the main chain can be achieved in one of two ways:

1. The polymer itself is made up of mesogenic monomer units which allow for restricted rotation between the units, resulting in a tendency to a rigid or semi-rigid rod conformation.
2. The polymer has a flexible backbone, but strong interactions between nearby monomers cause a transition from a random coil state to a rigid **helix**.

As an example of a polymer in the first category, Fig. 7.14 shows the chemical structure of an aromatic polyamide, poly(*p*-phenyleneterephthalamide) or PPTA, which forms a liquid crystal phase in solution. If fibres are spun from this liquid crystalline solution, they will have a high degree of molecular alignment resulting in very good mechanical properties; this material in fact forms the basis for the commercial material Kevlar (du Pont), which is very stiff and strong.

7.7.2   Helix–coil transitions

The classic case of a polymer which adopts a helical rigid rod structure stabilised by strong interactions between nearby monomers is provided by the α-**helix** structure of polypeptides. This is illustrated in Fig. 7.15; a hydrogen bond is formed between the C=O group on the ith monomer and the N–H group on the $(i+4)$th monomer.

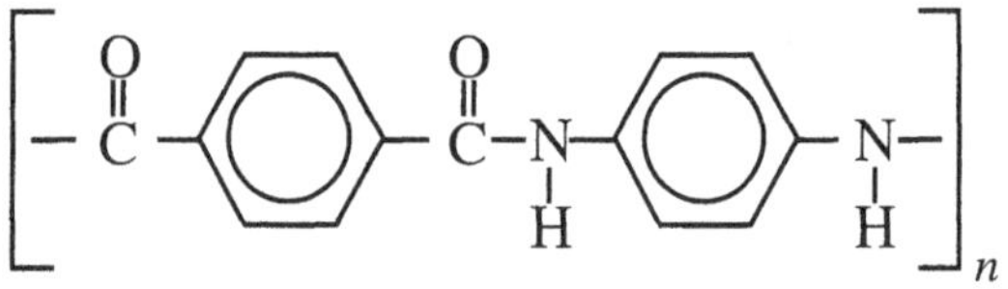

Fig. 7.14 The chemical structure of poly (*p*-phenyleneterephthalamide), or PPTA, a typical semi-rigid polymer.

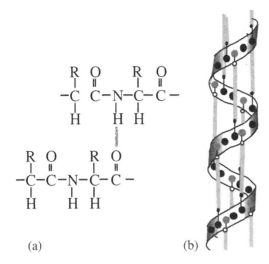

R O   R O
| ‖ | ‖
−C C−N−C− C −
| | |
H   H H

R O   R O
| ‖ | ‖
−C−C−N−C− C −
| | |
H   H H

(a)

(b)

Fig. 7.15 The formation of an α-helix in a polypeptide. (a) The chemical structure of a polypeptide, illustrating the hydrogen bond that can be formed between nearby monomers. (b) The α-helix that results when hydrogen bonds are formed between the ith and $(i+3)$th monomer units.

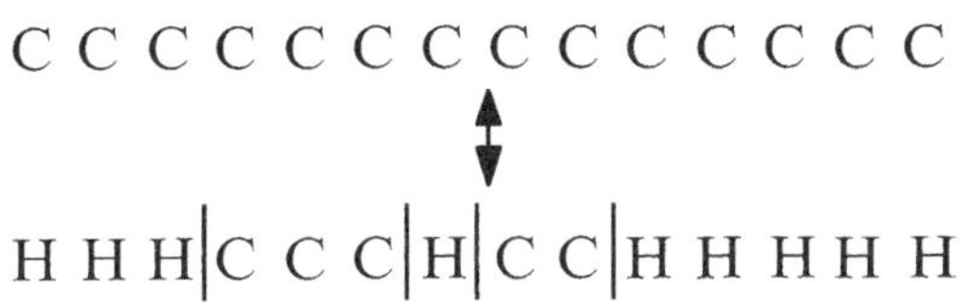

C C C C C C C C C C C C C C

⇕

H H H|C C C|H|C C|H H H H H

Fig. 7.16 A simple model for the helix–coil transition. We imagine our chain to be made up of units that are in either the **helix** state (H) or the **coil** state (C). The free energy change when a single specified coil unit transforms to a helix state is ΔF_{hc}; in addition there is a free energy associated with each junction between helix and coil segments ΔF_g.

What drives the formation of such a helical structure is the lowering of free energy that results from the formation of the hydrogen bonds. What opposes it is the loss of entropy that is a result of the loss of flexibility of the chain. Thus we would expect a transition to take place as the temperature is increased, with the lower energy helix conformation favoured at low temperatures, and the higher entropy coil conformation favoured at higher temperatures. What is the nature of this transition?

We can investigate this by making a simple but effective model, illustrated in Fig. 7.16. We imagine the chain to be made up of units that can be either in the helix state or in the coil state. For each specified unit, there is a certain free energy change ΔF_{hc} when it changes from a coil to a helix state; this represents the energy of the hydrogen bond which stabilises the helix. However, there is also a free energy ΔF_g associated with the junction between helix and coil segments. It is this factor that expresses the **cooperativity** of the transition. If one hydrogen bond has been formed, it makes it easier to form neighbouring hydrogen bonds.

To find out the nature of the transition, we need to determine the free energy change on going from an all-coil state to a state with a certain number of helix segments h, and a certain number of junctions between helix and coil regions $2g$. In addition to the free energy change associated with the total number of helices and the total number of junctions there is an entropy associated with the total number of different ways of arranging a chain with N units so that it has h helix segments and $2g$ junctions.

How can we find this entropy? The $2g$ junctions define a set of g boxes in which we have to put the helix segments, with at least one segment going in each box, and another set of g boxes into which the coil segments must be put. So for the helix segments, we define as Ω_h the number of ways there are of arranging h objects in g boxes, each of which must contain at least one object. This is given by

$$\Omega_h = \frac{h!}{g!(h-g)!}. \tag{7.17}$$

Similarly, for the coil segments there are Ω_c ways of arranging the $N - h$ coil segments in the g boxes, where

$$\Omega_c = \frac{(N-h)!}{g!(N-h-g)!}. \tag{7.18}$$

The total number of arrangements is $\Omega_c \times \Omega_h$, and the entropy associated with this number of arrangements is

$$\Delta S_c(h, g, N) = k_B \ln(\Omega_c \Omega_h). \tag{7.19}$$

We can simplify this using Stirling's approximation for the factorials ($\ln x! \approx x \ln x - x$), to find

$$\begin{aligned}\Delta S_c(h, g, N)/k_B = {} & h \ln h + (N - h)\ln(N - h) - 2g \ln g \\ & - (h - g)\ln(h - g) \\ & - (N - h - g)\ln(N - h - g).\end{aligned} \tag{7.20}$$

Now we can write down the free energy relative to the coil state in terms of h and g, $F(h, g)$;

$$F(h, g) = h \Delta F_{hc} + 2g \Delta F_g - T \Delta S_c(h, g, N). \tag{7.21}$$

We need to minimise $F(h, g)$ with respect to both h and g. Taking partial derivatives with respect to each variable and setting these equal to zero gives us two equations:

$$\frac{c(h-g)}{h(c-g)} = \exp\left(-\frac{\Delta F_{hc}}{k_B T}\right) = s, \tag{7.22}$$

$$\frac{g^2}{(c-g)(h-g)} = \exp\left(-\frac{2\Delta F_g}{k_B T}\right) = \sigma. \tag{7.23}$$

Here we have written the number of coil segments $N - h = c$, and we have introduced the two parameters σ and s. The parameter s expresses the preference of a given segment for the helix state over the coil state; as the temperature changes s may change from a value less than unity, in which the coil state is energetically preferred, to a value greater than unity, in which the helix state is preferred. The parameter σ is a measure of the cooperativity of the transition. For $\sigma = 1$ there is no cooperativity; the conformational state of one segment is independent of its neighbour. As the energy associated with a junction between helical and coil regions increases, σ takes a value less than unity, with a value of $\sigma = 0$ implying that junctions between helical regions and coil regions are absolutely forbidden.

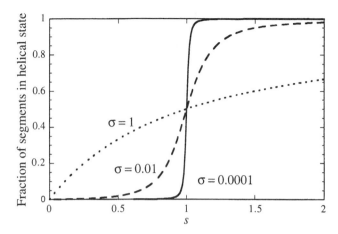

Fig. 7.17 The helix–coil transition as predicted by eqn 7.24. The coil state is favoured for $s < 1$, and the helix state for $s > 1$. For a system with no cooperativity ($\sigma = 1$), the transition is very broad, but for realistic values of the cooperativity parameter $\sigma \approx 10^{-4}$, the transition is rather sharp.

Writing $f_h = h/N$ for the fraction of segments in the coil state, we can eliminate g from the simultaneous equations 7.23 to find

$$f_h = \frac{1}{2} + \frac{(s-1)}{2\sqrt{(s-1)^2 + 4s\sigma}}. \tag{7.24}$$

In Fig. 7.17 we plot eqn 7.24 as a function of s for various values of σ. In all cases we see that for $s < 1$, for which the coil state is energetically favourable, the molecule is predominantly in the coil state, while for $s > 1$ the molecular is predominantly helical. But the nature of the transition between the coil and helix states depends on the value of σ. For large values of σ, the transition is very broad. For $\sigma = 1$, which corresponds to a value of $\Delta f_g = 0$, eqn 7.24 reduces to $f_h = s/(1+s)$. In this limit the probability of a segment being in the helix state is given by a simple Boltzmann factor; because there is no energy penalty for junctions between coil and helix segments each segment behaves independently. As the degree of cooperativity becomes larger, corresponding to smaller values of σ, the transition becomes much sharper.

The sharpness of the transition (values of σ are found to be in the range 10^{-4}–10^{-3}) makes it tempting to think of the helix–coil transition as being analogous to **melting**. However, there is an important fundamental difference. The helix–coil transition is not a true phase transition, as the width of the transition does not tend to zero as the total size of the system becomes infinite, but instead remains finite. This is a manifestation of a general theorem in statistical physics that states that true phase transitions do not occur in one-dimensional systems in which the interactions are of short range. For any finite value of s there will always be a finite fraction of coil sections, so a long chain will always have some shorter coil sections coexisting with longer sections of helix. Fully rigid rods will only be obtained for relatively short chains.

7.7.3   The isotropic/nematic transition for ideal hard rods

An important idealisation of a lyotropic liquid crystal system consists of a solution of **hard rods**, the theory of which was developed by Onsager. This is the anisotropic analogue of the hard-sphere system we discussed in Chapter 4, where a transition to an ordered crystalline phase was driven by an effective repulsion of entropic origin arising from the effect of excluded volume. The energy of interaction between hard rods is zero except when they overlap in space, in which case it is infinite. This means that there is a reduction in translational entropy, as some space is not available to be explored by a given rod because it is already occupied by another rod. In the hard-rod system, we find that less volume is excluded if the rods tend to align, and it is this reduction of the excluded volume effect by alignment which drives a phase transition from an isotropic state to a nematic liquid crystalline state as a function of increasing concentration of the rods.

Let us recall some basic results about excluded volume from Chapter 4. In a perfect gas one can write the entropy per atom S_{ideal} of N atoms in a volume V as

$$S_{ideal} = k_B \ln \left(a \frac{V}{N} \right) \tag{7.25}$$

where a is a constant. If the gas atoms have a finite volume b this reduces the volume accessible to any given atom from V to $V - Nb$, and the entropy is modified to

$$S = k_B \ln \left(a \frac{(V - Nb)}{N} \right)$$
$$= S_{ideal} + k_B \ln \left(1 - \frac{bN}{V} \right)$$
$$= S_{ideal} - k_B \left(\frac{N}{V} \right) b, \tag{7.26}$$

where we have expanded the logarithm assuming that the volume fraction of atoms is low. The corresponding free energy is

$$F = F_{ideal} + k_B T \left(\frac{N}{V} \right) b$$
$$= F_0 + k_B T \log c + k_B T c b, \tag{7.27}$$

where the concentration $c = N/V$.

In the hard-rod system another factor enters: the degree of alignment of the rods. If we describe this by an orientation function $f(\theta)$, we need to introduce two new factors to the expression for the free energy:

1. As a net orientation is introduced, there is a loss of orientational entropy ΔS, which is given by eqn 7.3 as $\Delta S = -k_B \int f(\theta) \ln[4\pi f(\theta)] \, d\Omega$.
2. The excluded volume b becomes a function of the degree of orientation; as the degree of alignment increases the excluded volume decreases.

To quantify the second point, consider two rods, each of length L and diameter D, that make an angle γ with each other. The excluded volume is $2L^2D|\sin\gamma|$, as shown in Fig. 7.18. If the orientation of all the rods is completely random, then the average value of $|\sin\gamma|$ can be shown to be $\pi/4$. However, as the rods become aligned then the average value of $|\sin\gamma|$ starts to decrease from this value. If we write this average value as $p[f(\theta)] = \langle|\sin\gamma|\rangle$ then in terms of the orientation function $f(\theta)$ we have

$$p[f(\theta)] = \langle|\sin\gamma|\rangle = \int\int f(\theta)f(\theta')\sin\gamma\,\mathrm{d}\Omega\,\mathrm{d}\Omega'. \tag{7.28}$$

Thus for any given distribution function one can calculate $p[f(\theta)]$ in terms of this rather messy multiple integral.

Taking together the term for the loss of orientational entropy and the excluded volume, we find for the free energy

$$F = F_0 + k_BT\left(\log(c) + \int f(\theta)\ln[4\pi f(\theta)]\,\mathrm{d}\Omega + L^2Dcp[f(\theta)]\right). \tag{7.29}$$

It is convenient to rewrite this in terms of the volume fraction of the rods $\phi = c\pi LD^2/4$; after absorbing more constant terms in F_0 this gives us

$$F = F_0' + k_BT\left[\log\left(\frac{L}{D}\phi\right) + \int f(\theta)\ln[4\pi f(\theta)]\,\mathrm{d}\Omega + \frac{4}{\pi}\frac{L}{D}\phi p[f(\theta)]\right]. \tag{7.30}$$

From this we can see that the phase diagram must be a function only of the combination $\phi L/D$, the product of the **volume fraction** and the **aspect ratio** L/D of the rods.

This expression is a **functional**, a function of a function, so to find the function $f(\theta)$ that minimises it soon leads us into rather difficult mathematics.[1] We can, however, get a good approximate solution by assuming a particular trial functional form for the distribution function, and then minimising the free energy with respect to a parameter in our guessed distribution function (this amounts in mathematical terms to a variational method of approximate solution).

A convenient, properly normalised, trial function is

$$f(\theta) = \frac{\alpha}{4\pi}\frac{\cosh(\alpha\cos(\theta))}{\sinh\alpha}. \tag{7.31}$$

Here the parameter α controls the degree of orientation: for $\alpha = 0$ the distribution is uniform, and as α increases the function develops sharp peaks around the directions $\theta = 0$ and $\theta = \pi$, giving shapes similar to those shown in Fig. 7.4. Once again, we can characterise the degree of order in terms of the order parameter $\mathcal{S} = \int \frac{1}{2}\left(3\cos^2\theta - 1\right)f(\theta)\,\mathrm{d}\Omega$.

Using this function we can now use eqn 7.28 to evaluate the average value of $|\sin\gamma|$ as a function of the degree of orientation of the rods. This is shown in Fig. 7.19; for completely randomly oriented rods it takes the value $\pi/4$, and then as the rods get more aligned its value falls. We can similarly calculate the loss of orientational entropy as a function of the order parameter, and thus we can plot the contributions to the free energy. These are sketched in Fig. 7.20.

This plot summarises the physics of the hard-rod system. As the degree of ordering increases, there is an *increase* in free energy associated with the loss

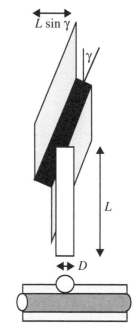

Fig. 7.18 Excluded volume in the interaction of two rods, each of length L and diameter D, with an angle γ between them. The presence of one rod means that a volume $DL^2|\sin\gamma|$ is inaccessible to the other rod.

[1] In fact it leads to an integral equation that needs to be solved numerically.

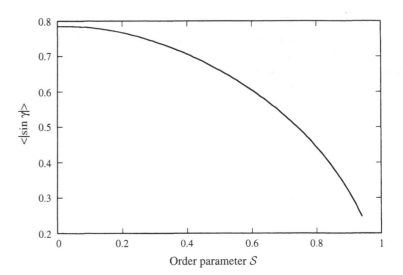

Fig. 7.19 The average value of the magnitude of the angle between two rods as a function of their degree of orientation.

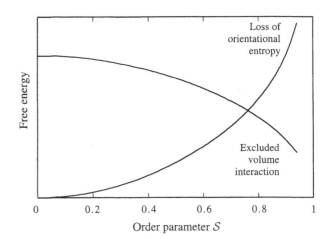

Fig. 7.20 Contributions to the free energy of a solution of hard rods as a function of their degree of orientation, as measured by the order parameter S. The relative importance of the two contributions depends on the product of the volume fraction of the rods and their aspect ratio.

of orientational entropy, but there is a *decrease* in the free energy arising from excluded volume interactions. The relative importance of the two terms depends on the quantity $\phi D/L$, the product of the volume fraction of the rods and their aspect ratio: as the volume fraction of the rods gets higher, or as the rods become more elongated, the excluded volume term becomes more important, and this can drive a phase transition into the nematic state.

This is illustrated in Fig. 7.21. For the lower value of the product $\phi L/D$, the free minimum energy state clearly occurs for $S = 0$; the equilibrium phase is an isotropic solution. But as this ratio is increased, a minimum appears in the free energy curve at a non-zero value of the order parameter S; a nematic phase has appeared. Analysis of the free energy curves reveals that there is a first-order phase transition from the isotropic to the nematic state. For values of $\phi L/D < 3.34$, the solution is isotropic. For values of $\phi L/D < 4.49$, the

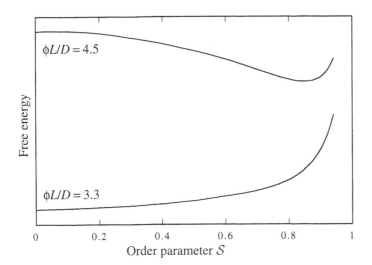

Fig. 7.21 The free energy of a solution of hard rods as a function of their degree of orientation, as measured by the order parameter S, for two values of the product of volume fraction and aspect ratio. For $\phi L/D = 3.3$, the lowest energy state is isotropic, but for the higher value of $\phi L/D = 4.5$ the equilibrium state has nematic order.

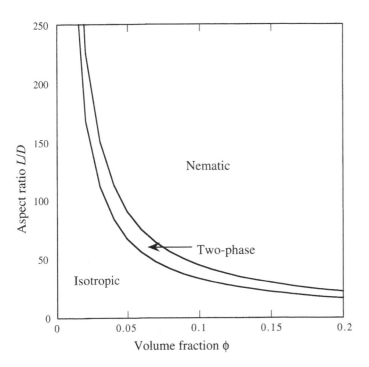

Fig. 7.22 The phase diagram of a solution of hard rods as a function of their volume fraction ϕ and aspect ratio L/D.

solution is nematic, while for values of $\phi L/D$ intermediate between these two limits, the solution separates into coexisting nematic and isotropic phases, with the order parameter in the nematic phase taking the value $S = 0.84$. The resulting phase diagram is shown as a function of ϕ and L/D in Fig. 7.22. One should note from this diagram that in order to obtain a liquid crystalline phase at a low volume fraction it is necessary to have rather a large aspect ratio.

7.7.4   Transitions in real lyotropic systems

Real polymer systems of the kind that form lyotropic phases differ from the ideal hard-rod systems discussed above in two ways:

1. Additional interactions, either attractive or repulsive, exist between the rods over and above the simple excluded volume interaction.
2. The rods may not be completely stiff.

These two factors can make the observed phase diagrams in real lyotropic systems considerably richer than those predicted from the theory of ideal hard rods.

The effect of rod–rod interactions is perhaps the most important. One can represent the effect of these interactions by an **interaction parameter** χ of the kind introduced in Chapter 3; this parameter represents in a dimensionless way the difference in energy between rod–rod, solvent–solvent, and rod–solvent interactions. Negative values of χ correspond to net repulsive interactions between the rods, while positive values of χ correspond to net attractive interactions. If they are strong enough, these attractive interactions can lead to liquid–liquid phase separation.

The resulting phase diagram can be calculated by different theoretical approaches which are beyond the scope of this treatment. We can, however, understand this phase diagram at a qualitative level. Figure 7.23 illustrates a phase diagram, calculated using a lattice theory due to Flory (see de Gennes and Prost 1993), for rods with an aspect ratio of 100. For repulsive interactions there is a narrow 'chimney' of two-phase coexistence; in this regime the behaviour is dominated by the excluded volume interaction. For attractive interactions (which normally become more important as the temperature is lowered) there is a region of what is essentially liquid–liquid phase separation, with the high volume fraction phase being nematic. There is a very narrow window of interaction parameters in which two nematic phases at different volume fractions coexist.

As the aspect ratio of the rods decreases, we would expect on the basis of our discussion of ideal hard rods that the volume fractions defining the two-phase chimney increase.

The other important complication in practical systems arises from the fact that physical systems are not perfectly stiff. Such chains are often referred to as **semi-flexible**. By this we mean that the statistical step length introduced in the discussion of random walks in Chapter 5 is significantly larger than the diameter of the chain, but still considerably less than the total length of the chain. In this case the phase diagrams are qualitatively similar to that shown in Fig. 7.22, but the volume fractions required to enter the nematic phase are considerably higher than in the rigid rod case and the degree of ordering in the nematic phase is somewhat less.

7.7.5   Thermotropic liquid crystal phases

Although a rigid or semi-rigid polymer such as PPTA (Fig. 7.14) forms liquid crystal phases in solution, it is difficult to obtain a liquid crystalline melt of the pure polymer. The melting point of the crystalline material is too high, and if heated the material will thermally decompose before the melting point is

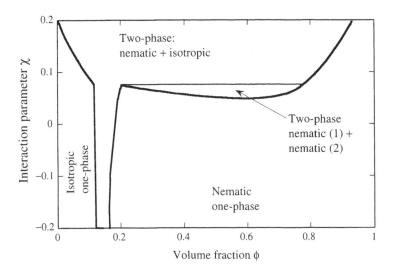

Fig. 7.23 The phase diagram of a solution of rigid rod-like polymers with an aspect ratio $L/D = 100$ as a function of their volume fraction ϕ and interaction parameter χ, as calculated from Flory's theory. Note that we would normally expect χ to decrease as the temperature is increased, so the corresponding diagram as a function of temperature would look inverted.

reached. To obtain a polymer that can be processed in the liquid crystalline melt state, one needs to be able to reduce the transition temperature from the crystal to the liquid crystal to an experimentally accessible value. Such materials are known as **thermotropic liquid crystalline polymers**. Among the strategies that can be employed to achieve this are:

- Inclusion of flexible units in the backbone. If the rigid units are separated by, for example, a $-(CH_2)_n-$ group, then the transition temperature from the crystal to liquid crystal phase is decreased as a function of increasing n.
- Use of a random copolymer. If two rigid units are copolymerised in a random way, the **quenched disorder** along the chain backbone will suppress the formation of a crystalline phase, while still permitting the formation of a liquid crystalline phase.
- Attachment of large, flexible side groups to the backbone. The melting point of a rigid polymer can be effectively reduced by attaching flexible side chains—typically aliphatic groups—to the backbone; the side chains can be thought of as acting like chemically attached solvent molecules. These materials are sometimes referred to as **hairy-rod** polymers. This approach also has the useful property that one can modify the solubility of such polymers in different solvents by modifying the chemical nature of the side groups. For an example of such a material, see Fig. 7.24.

Thermotropic liquid crystalline polymers have been explored commercially as high-specification structural materials; if they can be processed in the liquid crystalline state the high degree of molecular alignment leads to extremely high strength and stiffness. Another more recent application occurs in the field of semiconducting polymers. Such polymers have a conjugated backbone, which renders them insoluble and very difficult to process unless side groups can be attached, as in the example shown in Fig. 7.24. If, as a result of the attachment of such side groups, the materials show liquid crystalline phase behaviour then one can hope to process them in a way which maximises the degree of molecular order in the final electronic or optoelectronic device, and

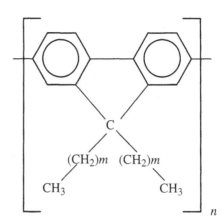

Fig. 7.24 The chemical structure of a polyfluorene, a class of conjugated polymers with a rigid rod backbone, which exhibits thermotropic liquid crystalline behaviour when aliphatic side groups of moderate length are attached (typically $m = 7$).

thus optimises the device parameters such as the charge carrier mobility or the degree of polarisation of emission.

Further reading

The theory of small-molecule liquid crystals is covered in de Gennes and Prost (1993); there is also much useful material in Chaikin and Lubensky (1995). Chandrasekhar (1992) is a more general monograph, covering theory, experiment, and the fundamentals of device applications. Polymer liquid crystals are described in Donald and Windle (1992).

Exercises

(7.1) In a simple model of a nematic/isotropic phase transition, the free energy change ΔF on going from the isotropic state to an ordered state with an order parameter S can be written as a function of temperature T as

$$\Delta F = \frac{1}{2}a(T - T^*)S^2 - wS^3 + uS^4,$$

where a, u, w, and T^* are positive constants.

a) Sketch curves of ΔF as a function of S for various values of temperature T for the following parameters: $a = 0.0033$; $T^* = 300$; $w = 0$; $u = 1$. What is the nature of the phase transition at $T = T^*$?

b) Sketch curves of ΔF as a function of S for temperatures T in the range 300–500 for the following parameters: $a = 0.0033$; $T^* = 300$; $w = 1$; $u = 1$. What is the nature of the phase transition now?

c) Show that for this model the nematic/isotropic transition occurs at a temperature T_c given by

$$T_c = T^* + \frac{w^2}{2au}.$$

d) Derive an expression for the order parameter at the transition S_c.

(7.2) You are asked to design a twisted nematic display, using a nematic liquid crystal whose elastic constants are given by $K_1 = 5.3 \times 10^{-12}$ N, $K_2 = 2.2 \times 10^{-12}$ N, and $K_3 = 7.45 \times 10^{-12}$ N. If the dielectric anisotropy $\epsilon_a = 0.7\epsilon_0$, what is the switching voltage?

(7.3) A biopolymer is observed to change from a helix to a coil state over a 5 K temperature interval centred on 343 K.

a) Using eqn 7.24, show that the width of the helix-coil transition in terms of the parameter s may be characterised by $\Delta s = 4\sigma^{1/2}$.

b) Assuming that s is a linear function of temperature close to the midpoint of the transition, estimate the free energy, in units of $k_B T$, associated with a junction between helical and coil sections, $\Delta F_g / k_B T$.

c) How would you expect the width of the transition as the length of the biopolymer is increased? What is the significance of this result?

Molecular order in soft condensed matter—crystallinity in polymers

8

8.1 Introduction

The most common state of molecular order in polymers is the crystalline state, with full three-dimensional positional order. However, in contrast to the situation in elemental solids and small molecules, very few polymer systems can attain a state of complete crystallinity. Instead, almost all polymers are **semi-crystalline**, consisting of a composite of very small crystals in a matrix of much less ordered material, with a total of between 20% and 60% of the material being present in the crystalline state. The amorphous material can be either glassy (as for example in polyethylene terephthalate, familiar from bottles for carbonated soft drinks) or rubbery, that is to say liquid-like, but effectively cross-linked by the small crystals. This is the situation in polyethylene. The reasons for this partial crystallinity are:

- **Slow kinetics.** Polymers are entangled, so it takes a long time for the molecules to arrange themselves in perfect crystals. Even quite modest cooling rates allow one to produce a glass.
- **Quenched disorder.** Polymers may have disorder built into the polymer chain. This can take the form of randomness in the stereochemistry or tacticity (polystyrene for example is usually atactic and does not readily crystallise at all), or a random sequence of monomers in a random copolymer.
- **Branching.** If the polymer chains have many branches, this makes it more difficult to pack the chains into regular crystals. This is why low-density polyethylene, which is branched, is less crystalline than high-density polyethylene, which is strictly linear.

8.2 Hierarchies of structure

Crystalline polymers have a hierarchical structure, that is there are different structures on different length scales.

The basic unit of most polymer crystals is the **chain-folded lamella** (Fig. 8.1). The lamellar thickness l is independent of molecular weight; as we shall see it is a function of the supercooling when the crystal was formed. A typical value for l would be about 10 nm. Lamellae are separated by amorphous regions;

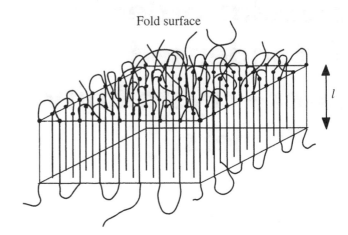

Fold surface

Fig. 8.1 A chain-folded lamella, the basic unit in semi-crystalline polymers. The lamellar thickness l is much smaller than the contour length of a polymer chain.

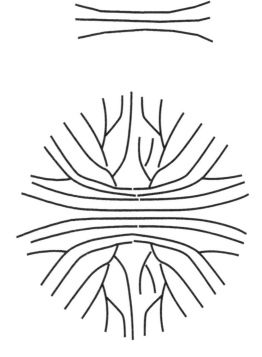

Fig. 8.2 The assembly of a **spherulite** from chain-folded lamellae radiating and branching from a central nucleus.

individual chains may be involved in more than one lamella as well as the amorphous regions in between.

The chain-folded lamellae are themselves organised in larger scale structures called **spherulites** (Fig. 8.2). These structures, which may be many micrometres in size (see e.g. Fig. 8.3), consist of a sheaves of individual lamellae which grow out from a central nucleus, until finally the whole of space is filled by these structures.

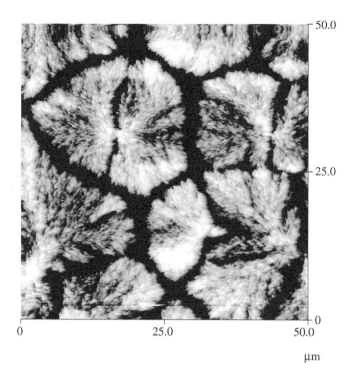

50.0

25.0

0

0   25.0   50.0

µm

Fig. 8.3 Spherulites in thin films of nylon, imaged using atomic force microscopy. Image courtesy of D. Jones.

8.3   Chain-folded crystals

Why is the chain-folded lamellar crystal the basic unit of semi-crystalline polymers? These are not **equilibrium** structures; the fold surface is considerably disordered, and there is a substantial interfacial energy σ_f associated with it. This means that the melting point of a lamellar crystal of thickness l, $T_m(l)$, is depressed from the ideal thermodynamic value, $T_m(\infty)$. This can be seen by considering the change in free energy δg when one polymer stem, of length l, joins the crystal. If the cross-sectional area of the stem is a^2, then we can write this change in energy as

$$\Delta g = -\frac{\Delta H_m \Delta T}{T_m(\infty)} l a^2 + 2a^2 \sigma_f, \qquad (8.1)$$

where ΔH_m is the latent heat of melting per unit volume, and the undercooling $\Delta T = T_m(\infty) - T$. So the melting point of this crystal of finite thickness is given by the condition $\Delta g = 0$, which yields the expression

$$T_m(l) = T_m(\infty) \left(1 - \frac{2\sigma_f}{\Delta H_m L} \right). \qquad (8.2)$$

Thus at equilibrium we would anticipate crystals in which the participating chains were fully extended. In fact, it turns out that the finite thickness of chain-folded polymer lamellar crystals arises from kinetic considerations, not equilibrium thermodynamics. In short, polymers form lamellae of a well-defined thickness because crystals with this thickness grow the fastest.

Where does this mechanism of length scale selection come from? The important insight is that in order for a section of a polymer chain to join the

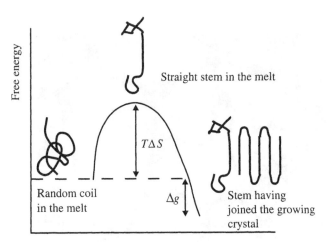

Fig. 8.4 Free energy changes when a stem of polymer joins the growing crystal from the melt.

crystal as a linear stem of length l, it has first to **unfold**; this requires a temporary loss of entropy, whose probability decreases exponentially with the length of the stem required. Thus we can see that:

- crystals that are too thick grow slowly because the probability of a long enough length of polymer straightening itself out from its random coil configuration is too small, while
- crystals that are too thin involve too much unfavourable energy from their fold surfaces, leading to smaller thermodynamic driving forces for the growth of the crystal.

We can make this argument more quantitative with a simple model calculation. The energetics involved when one stem is added to the crystal are sketched in Fig. 8.4. In order to join the crystal, a length of chain, which in the melt has a random coil configuration, must first straighten itself out. This leads to a reduction in entropy ΔS. Only when a random fluctuation has produced such a straight length of chain can the stem join the growing crystal face, finishing up with a free energy Δg lower than its energy in the melt.

Now we can estimate both the rate at which chain segments join the growing crystal from the melt, and the rate at which segments leave the crystal to rejoin the melt. We have

$$\text{melt} \to \text{crystal rate} = \tau^{-1} \exp\left(-\frac{\Delta S}{k_{\mathrm{B}}}\right)$$

$$\text{crystal} \to \text{melt rate} = \tau^{-1} \exp\left(-\frac{(T\Delta S - \Delta g)}{k_{\mathrm{B}}T}\right), \tag{8.3}$$

where τ^{-1} is a microscopic frequency (recall also that in eqn 8.1 we defined Δg to be negative when the crystal was stable). The difference between these two rates gives us the net crystallisation rate u, defined as the number of stems attached to a given site per unit time:

$$u = \tau^{-1} \exp\left(-\frac{\Delta S}{k_{\mathrm{B}}}\right)\left[1 - \exp\left(\frac{\Delta g}{k_{\mathrm{B}}T}\right)\right]. \tag{8.4}$$

To simplify things we will assume that $\Delta g/k_{\mathrm{B}}T$ is small enough to expand the exponential. Writing for the velocity of crystal growth $v = ua$ (where a is the cross-sectional diameter of the polymer chain), we find

$$v = a\tau^{-1} \exp\left(-\frac{\Delta S}{k_B}\right) \frac{\Delta g}{k_B T}. \tag{8.5}$$

We now need to see how this growth velocity depends on the thickness of the crystal l. Equation 8.1 gives us the variation of Δg on l. The entropy loss ΔS on straightening out a length l of the chain is proportional to the number of segments in the length to be straightened. We write $\Delta S = \mu l/a$ where μ is a dimensionless constant. This gives us

$$v(l) = (\text{constant}) \left(2a^2\sigma_f - \frac{\Delta H_m \Delta T}{T_m(\infty)} l_f a^2\right) \exp\left(-\frac{\mu l}{a}\right). \tag{8.6}$$

The shape of this function is plotted in Fig. 8.5. Crystals of thickness l_c are at equilibrium with the melt at a given temperature and do not grow at all. On the other hand there is a certain thickness l^* for which the growth rate is a maximum, and it is crystals of this thickness which we expect to dominate the final morphology. We can find l^* by differentiating eqn 8.6; this yields

$$l^* = \frac{a}{\mu} + \frac{2\sigma_f T_m(\infty)}{\Delta H_m(T_m(\infty) - T)}. \tag{8.7}$$

We see that the deeper the quench the thinner the resulting crystals will be. This is what is observed experimentally, and indeed the functional form of eqn 8.7 gives a good fit to the data, as shown in Fig. 8.6.

We can also use this approach to predict the temperature dependence of the crystal growth rate. To do this we simply substitute our expression for the fastest growing crystal thickness l^* from eqn 8.7 into our expression for the crystal growth velocity, eqn 8.5. An important point that we have up to now glossed over is the question of what determines the microscopic frequency τ^{-1}. This gives a measure of the rate of conformational rearrangements of a polymer coil in a melt. These conformational rearrangements involve the complicated internal dynamics of a polymer chain. Luckily, however, even in the absence of a detailed analysis of these modes we know from the principle

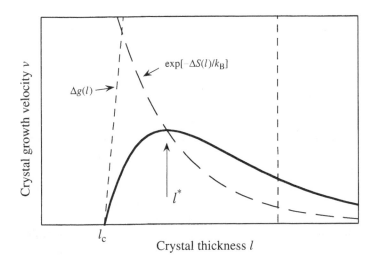

Fig. 8.5 The crystal growth rate as a function of crystal thickness. At a given temperature, crystals of thickness l_c are in equilibrium with the melt and do not grow at all. Crystals of thickness l^* have the maximum growth rate and dominate the morphology.

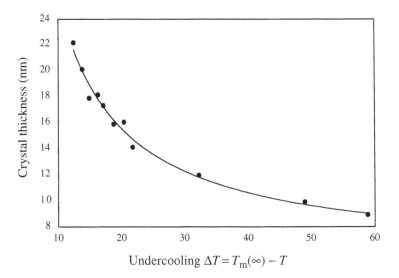

Fig. 8.6 Crystal thickness as a function of undercooling for polyethylene, showing good agreement with the functional form of eqn 8.7. Data from Barham *et al.* (1985).

of time–temperature superposition (see Section 5.5.3) that all such modes have the same temperature dependence, which we can write in the Vogel–Fulcher form. Thus we can write

$$\tau^{-1} = \tau_0^{-1} \exp\left(\frac{-B}{T - T_0}\right),$$ (8.8)

where B is a positive coefficient and T_0 is the Vogel–Fulcher temperature, a material constant which is typically around 50°C below the experimental glass transition temperature.

Putting all this together, we find for the temperature dependence of the crystal growth velocity

$$v = \frac{a\tau_0^{-1}}{e\,k_\mathrm{B}T}\frac{a^3}{\mu}\frac{\Delta H_\mathrm{m}\Delta T}{T_\mathrm{m}(\infty)}\exp\left(\frac{-B}{T - T_0}\right)\exp\left(-\frac{2\mu\sigma_\mathrm{f}T_\mathrm{m}(\infty)}{a\Delta H_\mathrm{m}\Delta T}\right).$$ (8.9)

The temperature dependence of this expression is dominated by the two exponentials, which combine to give a very strong peak in the growth rate as a function of temperature. At high temperatures, it is the size of the thermodynamic driving force that controls the growth rate, via the second exponential function in eqn 8.9. At lower temperatures, it is the rapidly decreasing mobility as the glass transition approaches that suppresses the growth of crystals. In practice, it is not at all uncommon to observe changes in crystal growth rates of three or four orders of magnitude, so great is the temperature dependence predicted by eqn 8.9.

To conclude this section, it is worth making some general observations. We have seen that when polymers crystallise, kinetic considerations mean that the final morphology is a long way from equilibrium, and indeed the equilibrium morphology, in the sense of the morphology with overall lowest free energy, is practically unattainable. If a semi-crystalline polymer is kept at a temperature at which the chains are still mobile, there is some evolution of the morphology which is reflected in a very slow increase in the crystalline fraction. The crystalline lamellae do thicken; new lamellae are nucleated and grow in the interstices between existing lamellae. But to achieve the equilibrium state of fully chain extended crystals would require a wholesale rearrangement of chains, which would involve such a huge energy barrier that it is practically unattainable. Here we see the limits of equilibrium thermodynamics.

Further reading

Many books about polymer physics describe semi-crystalline polymers; Strobl (1997) is particularly clear. For a monograph on polymer morphology, with particular emphasis on the experimental contributions of electron microscopy, see Bassett (1981).

Exercises

(8.1) The table below gives the measured values of the initial lamellar thickness for polyethylene crystalised at various temperatures. The equilibrium melting point is independently measured and is found to be 417.8 K.

Temperature/ K	358.95	368.95	385.75	396.15	397.55
Crystal thickness/ nm	8.9	9.9	12.0	14.1	16.1

Temperature/ K	399.15	400.85	401.65	403.05	404.15	405.55
Crystal thickness/ nm	15.9	17.3	18.2	17.9	20.1	22.2

a) Draw a graph of the data to test the relation eqn 8.7.
b) What is the melting point of crystals formed at 400 K?

(8.2) A monodisperse fraction of polyethylene is synthesised with the chemical formula $C_{122}H_{246}$. Use the relation you found from the data in the last question to predict the degree of undercooling needed for the initial lamellar thickness predicted by eqn 8.7 to correspond

a) to a crystal whose thickness is exactly equal to the fully stretched polyethylene chain,

b) to a crystal in which each chain is folded exactly once.

[Take the fully stretched length of a polyethylene chain with N carbon atoms to be $0.1275N$ nm.]

9 Supramolecular self-assembly in soft condensed matter

9.1 Introduction

One of the most spectacular properties of soft matter systems is their ability to **self-assemble** on a hierarchy of length scales: molecules arrange themselves in supramolecular assemblies, and these assemblies in turn pack in structures that may be highly ordered, even though the units that are being packed are non-crystalline and of dimensions considerably bigger than atomic scales. The most common examples of these self-assembled phases are found in soap and other materials that are formed by mixing water with a class of molecules known as **amphiphiles**, which have the property that part of the molecule has an affinity to water and another part of the molecule is repelled from water. When such molecules are dissolved in water, the way the system attempts to satisfy these two irreconcilable tendencies leads to a rich array of possible structures.

Many amphiphilic molecules are relatively small, but polymers too can display this tendency if they are composed of two or more chemically different blocks covalently linked together. In solution such **block copolymers** behave in very similar ways to small-molecule amphiphiles. However, the large size of polymer molecules has important consequences for the thermodynamics of mixing. Even when the chemical units making up a pair of polymers are relatively similar, and the unfavourable energies of interaction between them small, the entropy of mixing is usually not large enough to promote mixing and the polymers will phase separate. If the two polymers are linked together covalently, this constraint prevents them from phase separating on a macroscopic scale and the result is **microphase separation**, an ordered arrangement of domains of the two polymer types on mesoscopic length scales. This is another example of **self-assembly**—complex ordered structures arising solely from the requirement that the system minimises its free energy.

9.2 Self-assembled phases in solutions of amphiphilic molecules

9.2.1 Why oil and water do not mix

Oil and water provide the most familiar examples of liquids which do not mix. We can understand the behaviour of mixtures of water with hydrocarbons in terms of the framework established in Chapter 3. On taking one oil molecule from a reservoir of similar molecules and putting in an environment of water, there is an increase in free energy; if we try and mix oil and water the entropy we gain by mixing the two species is more than offset by this unfavourable interaction between the molecules and the lowest total free energy is found for a state in which the oil and water are found in macroscopically distinct phases.

In Chapter 3 we introduced a simple model to describe the statistical mechanics of mixing liquids: the regular solution model. We can use this as a framework for our discussion, but there is one important refinement that needs to be made. In the regular solution model, we write the free energy of mixing as the sum of two terms: one gives the gain in configurational entropy obtained in the mixed state, and the other was introduced as the energy of mixing, for which we derived a simple expression assuming simple pairwise interactions between neighbouring molecules. For mixtures involving water, the division between a configurational entropy term and a molecular interaction term is still valid, but the interaction is much more complicated than the balance of bond energies we used in Chapter 3. The character of the molecular interaction between water and the hydrocarbon-based molecules we collectively know as 'oils' is so important and so unusual that it has a special name—the **hydrophobic interaction**. In particular, it can be shown that the interaction has a significant entropic component that is quite distinct from the usual entropy of mixing.

To see this, we note that the free energy change at 298 K on taking one molecule of n-butane from pure butane and inserting it in pure water is 4.1×10^{-20} J. This is about $10 \, k_B T$; from this one can understand why hydrocarbons like butane are so insoluble in water. But if we measure this free energy change as a function of temperature, we can separate the contributions of entropy and enthalpy to this value, and we find that about 85% of this interaction arises from a decrease in entropy that occurs when a molecule of hydrocarbon is solubilised in water. How can the introduction of a foreign molecule into water decrease the entropy? The answer arises from the anomalous character of liquid water, which contains a network of strong hydrogen bonds. If we insert into this network a foreign molecule which cannot form hydrogen bonds, then the network locally rearranges to maximise the number of hydrogen bonds that can be formed. This restricts the number of configurations of the water molecules close to the solute, thus decreasing their contribution to the entropy. In short, the solute molecule forces the nearby water molecules to adopt a more ordered state in an attempt to satisfy their hydrogen bond requirements.

It is thus the hydrophobic interaction that leads to phase separation between oil and water. If we attempt to disperse oil and water, say by violent stirring or shaking, we will succeed in creating small particles of oil, but these are unstable and will rapidly coalesce until we recover two macroscopically distinct phases. Phase separation between oil and water takes place on a **macroscopic** length scale. We shall see later that certain more complicated molecules— amphiphiles—also tend to separate from water, driven by the hydrophobic interaction, but that in some cases this separation is limited in extent, resulting in the formation of thermodynamically stable aggregates of microscopic sizes. These aggregates are called **micelles**.

9.2.2 Aggregation and phase separation

We discussed the growth of large aggregates following liquid–liquid phase separation in Chapter 3. Here we discuss phase separation from a slightly different viewpoint, which will be useful when we come to analyse the self-assembly of micelles.

We begin by considering a solute with a tendency to phase-separate or aggregate, and we consider what proportion of the solute molecules at equilibrium will be present as isolated molecules, and what proportion will be in the form of aggregates containing some finite number of molecules. Suppose the overall volume fraction of solute is ϕ, and the volume fraction of solute molecules in an aggregate with N molecules is X_N. Clearly when we sum over all the possible aggregate sizes we recover the total volume fraction, so $\sum_N X_N = \phi$. The condition for equilibrium is that the chemical potential μ of the solute molecules is the same in all the different coexisting aggregates. This allows us to write

$$\mu = \epsilon_N + \frac{k_B T}{N} \log \frac{X_N}{N}. \tag{9.1}$$

Here ϵ_N is the free energy change when a specified molecule is taken from the bulk and put into an aggregate of N molecules. The second term represents the contribution arising from the translational entropy of the micelle. Note that the bigger the micelle, the smaller the importance of this translational entropy.

We can rearrange this equation to get an expression for X_N:

$$X_N = N \exp\left(\frac{N(\mu - \epsilon_N)}{k_B T}\right), \tag{9.2}$$

and if we put $N = 1$ we can rewrite this equation, eliminating μ, to derive an expression giving us the volume fraction of solute in the aggregate of size N in terms of the fraction of free solute molecules, X_1,

$$X_N = N \left[X_1 \exp \frac{(\epsilon_1 - \epsilon_N)}{k_B T} \right]^N. \tag{9.3}$$

The implications of this equation are straightforward. If $\epsilon_N \geq \epsilon_1$, then most of the solute molecules will be present in the solution as isolated molecules. On the other hand, if $\epsilon_N < \epsilon_1$, then aggregates will form. According to the specific way in which ϵ_N depends on N, the aggregates may be either finite or infinite in size.

To illustrate this point, consider the mixing of a pair of simple fluids. A spherical aggregate with N molecules will have a radius $r = (3Nv/4\pi)^{1/3}$, where v is a molecular volume, and there will be an energy $4\pi r^2 \gamma$ associated with the surface of this sphere, where γ is the interfacial energy. Thus we can write ϵ_N in terms of ϵ_∞ as

$$\epsilon_N = \epsilon_\infty + \frac{4\pi}{N}\left(\frac{3Nv}{4\pi}\right)^{2/3}$$

$$= \epsilon_\infty + \frac{\alpha k_B T}{N^{1/3}}, \tag{9.4}$$

where $\alpha k_B T = 4\pi\gamma(3v/4\pi)^{2/3}$. If we now use this expression for ϵ_1 and ϵ_N and substitute into eqn 9.3 we find

$$X_N = N\left\{ X_1 \exp\left[\alpha\left(1 - \frac{1}{N^{1/3}}\right)\right] \right\}^N \tag{9.5}$$

$$\approx N[X_1 \exp(\alpha)]^N. \tag{9.6}$$

This equation has a very straightforward interpretation. If the volume fraction of isolated solute molecules X_1 is small, such that $X_1 \exp(\alpha) < 1$, then very few larger aggregates are present. When X_1 approaches $\exp(-\alpha)$, then because all the volume fractions X_N must remain smaller than unity, if any more solute is added to the system it cannot remain as isolated molecules, but instead it must join an aggregate. Because ϵ_N is a monotonically decreasing function of N, then the size of the aggregate it joins is effectively infinite. There is thus a critical volume fraction, ϕ_c, above which isolated solute molecules at this volume fraction coexist with an infinite aggregate; this volume fraction is known as the **critical aggregation concentration** or CAC.

The fact that for simple fluids ϵ_N is a monotonically decreasing function of N means that the aggregates that appear above the CAC are effectively infinite in size. In more complex solutes—amphiphiles—we shall see that ϵ_N has a minimum for some finite value of N. In these circumstances, above the critical volume fraction the aggregates that appear are of finite size. These aggregates are known as micelles, and in these circumstances the critical volume fraction is known as the **critical micelle concentration** or CMC. In the next section we discuss why ϵ_N has a minimum for these molecules.

9.2.3 The aggregation of amphiphilic molecules

Amphiphiles are molecules containing chemical groups with widely differing affinities for water. A typical amphiphile, such as soap, consists of a hydrophilic group, sometimes ionic in character, attached to a hydrophobic section of the molecule, typically a hydrocarbon chain. Such molecules, when in aqueous solution, have a propensity to self-assemble to form remarkable structures. These structures include micelles, which may be spherical or cylindrical in shape, **bilayers**, and **vesicles**. Examples of some of these shapes are illustrated in Fig. 9.1. Moreover these aggregates, each of which contains many amphiphilic molecules, may themselves associate in ordered or disordered arrangements to produce remarkable complex phases.

Some examples of typical amphiphiles are given in Table 9.1. There is a wide variety of chemical structures; the hydrophilic head-group may be an uncharged

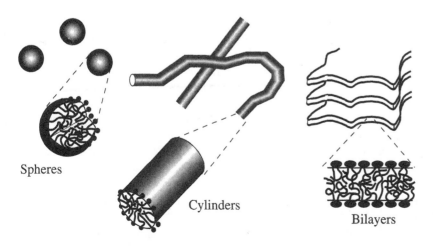

Spheres

Cylinders

Bilayers

Fig. 9.1 Types of aggregates that are encountered in amphiphile solutions.

Table 9.1 Examples of common amphiphiles

Example	Hydrophobic group	Hydrophilic group	Category	Where found
Sodium stearate	$C_{18}H_{37}$	$-COO^-Na^+$	Anionic	Soap
Sodium dodecyl sulphonate (SDS)	$C_{12}H_{25}$	$-OSO_3Na^+$	Anionic	Detergents
Hexadecyl trimethyl- ammonium bromide (CTAB)	$C_{16}H_{33}$	$-N^+(CH_3)_3Br^-$	Cationic	Mild disinfectants
$C_{12}E_5$	$C_{12}H_{25}$	$-(OCH_2CH_2)_5$	Non-ionic	Cosmetics
Pluronic P105	$(OCH_2C_2H_5)_{58}$	Two chains, each $-(OCH_2CH_2)_{37}$	Non-ionic, triblock copolymer	Cosmetics, pharmaceuticals
Lecithin	Two chains, each $C_{12}H_{25}$	Phosphatadyl choline	Zwitterionic, phospholipid	Animal cell membranes, food

group with an affinity for water, such as a sugar or a short polyether segment. More commonly the hydrophilic group is charged. In soaps and synthetic detergents the head-group is an anion; cationic head-groups produce molecules with anti-bacterial properties, which find use as mild antiseptic and disinfectant preparations. The hydrophobic tail of the molecule is almost universally made from aliphatic chains, which may be saturated or partially saturated. Soaps and synthetic detergents have a single hydrocarbon chain, but one very important class of amphiphiles has two such chains attached to its hydrophilic head-groups. These are the **phospholipids**, which are major components of biological membranes.

Despite the wide variety of chemical structures of amphiphiles, much of their behaviour can be understood in terms of a very simple model which captures most of the physics in terms of a few parameters which characterise the geometry of the molecule in rather a coarse way. The basic driving force behind the self-assembly process is the need to minimise the free energy, by minimising the degree of mixing between the hydrophobic tails of the amphiphile with water, while keeping the hydrophobic head-groups in contact with water. In the last section we introduced the quantity ϵ_N, as the free energy change on taking a molecule from the bulk and putting it into an aggregate with N other molecules. It is the balance between the need to keep the hydrophobic tails away from water and the hydrophilic groups in contact with water that leads to ϵ_N having a minimum for a certain well-defined value of N. This leads to a kind of thermodynamically arrested phase separation; rather than forming two macroscopically separated phases, as happens when we mix oil and water, microscopic aggregates are formed. The most common type of aggregate is the **spherical micelle**.

Some insight into the factors which determine what kind of micelle is formed by a given amphiphile is obtained from a simple geometrical argument due to Israelachvili. In this approach, we characterise an amphiphile by three parameters: the **optimum head-group area** a_0, the **critical chain length** l_c, and the **hydrocarbon volume** v. The hydrocarbon volume is simply the volume of the hydrocarbon chain, while the critical chain length is related to the length

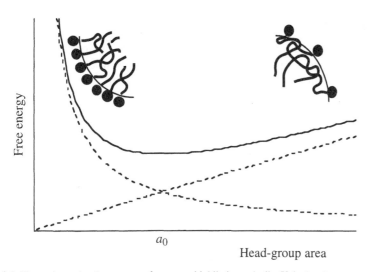

Fig. 9.2 The optimum head-group area for an amphiphile in a micelle. If the headgroups are too tightly packed, they repel each other by electrostatic and other forces, leading to an increase in free energy. If they are too far apart, hydrophobic chains are forced to come into contact with the water. The optimum head-group area a_0 is set by a balance between these two factors.

of the hydrocarbon chain if it is in a fully extended, straight configuration. The optimum head-group area needs more consideration.

The idea of an optimum head-group area for an amphiphile in a micelle is illustrated in Fig. 9.2. If the head-groups are forced too closely together they will repel each other by electrostatic and other interactions. On the other hand, if the head-groups are too far apart, this forces the hydrophobic tails to come into contact with the water, with a resulting increase in interfacial energy. A balance between these two factors gives rise to the optimum head-group area a_0. Note that this is not solely a geometric factor; because it arises from a balance between attractive and repulsive forces, if the range of the forces is modified (e.g. by changing the salt concentration to screen out electrostatic forces) then the optimum head-group area can be altered.

The condition for various shapes of micelle to be adopted can now be derived. For a sphere of radius r, if there are M molecules in the micelle the volume is $4\pi r^3/3 = Mv$ and the surface area $4\pi r^2 = Ma_0$. This means that the radius is $3v/a_0$. In order to be able to form a spherical micelle this radius must be less than the critical chain length l_c, giving the condition for spherical micelles as

$$\frac{v}{l_c a_0} \leq \frac{1}{3} \quad \text{(spheres)}. \tag{9.7}$$

For cylinders, given M molecules in length l of cylinder of radius r, the volume $\pi r^2 l = Mv$ and the surface area $2\pi r l = ma_0$. Thus the radius is $2v/a_0$, and we find the condition for cylindrical micelles is

$$\frac{1}{3} < \frac{v}{l_c a_0} \leq \frac{1}{2} \quad \text{(cylinders)}. \tag{9.8}$$

When the combination $v/(l_c a_0)$ becomes larger than 1/2, then the favoured shape of the aggregate will be a bilayer.

$N < M$

$N = M$

$N > M$

Fig. 9.3 The optimum aggregation number, M for a spherical micelle. Micelles containing a smaller number of molecules have too large an area for each head-group, causing unfavourable water/hydrocarbon interactions, while in micelles larger than the optimum number some head-groups are forced inside the hydrophobic core.

9.2.4 Spherical micelles and the CMC

If the packing geometry of an amphiphile favours the formation of spherical micelles, then we expect ϵ_N (the free energy change when a specified molecule is taken from the bulk and put into an aggregate of N molecules) to have a minimum value for a certain aggregation number $N = M$. This is illustrated in Fig. 9.3; micelles smaller than the optimum size will have too large an area per head-group, allowing energetically costly contacts between water and the hydrophobic core. Micelles larger than the optimum size will have their head-groups too closely packed together, and owing to the difficulty of packing the hydrocarbon chains at constant density some head-groups will be forced inside the hydrophobic core.

We can see the effect of the minimum in ϵ_N by assuming a simple quadratic form around the minimum; in this ansatz we put

$$\epsilon_N = \epsilon_M + \Lambda (N - M)^2. \tag{9.9}$$

We can use eqn 9.2 to obtain an expression relating X_N and X_M; this gives

$$X_N = N \left[\frac{X_M}{M} \exp\left(\frac{M(\epsilon_M - \epsilon_N)}{k_B T} \right) \right]^{N/M}. \tag{9.10}$$

Substituting in our ansatz for ϵ_N we find

$$X_N = N \left[\frac{X_M}{M} \exp\left(\frac{-M\Lambda(M - N)^2}{k_B T} \right) \right]^{N/M}. \tag{9.11}$$

For M large this gives a near-Gaussian distribution, with the variance in aggregation number given by

$$\langle |N - M|^2 \rangle = \frac{k_B T}{2M\Lambda}. \tag{9.12}$$

If the minimum is deep compared to $k_B T$, then the size distribution is rather narrow. It is instructive to examine the limit in which we consider only monomers and micelles with aggregation number M. Then we find

$$X_M = M \left[X_1 \exp\left(\frac{\Delta\epsilon}{k_B T} \right) \right]^M, \tag{9.13}$$

where $\Delta\epsilon = \epsilon_1 - \epsilon_M$. The total amphiphile volume fraction $\phi = X_1 + X_M$. From this equation we can see that for values of $X_1 < \exp(-\Delta\epsilon/k_B T)$ the amount of surfactant present in micelles is vanishingly small. The volume fraction $\phi_c = \exp(-\Delta\epsilon/k_B T)$ represents the CMC. The variation of the volume fraction in monomers and volume fraction in micelles is shown in Fig. 9.4.

9.2.5 Cylindrical micelles

Systems that form cylindrical micelles behave in a very different way to those that form spherical micelles. This is because ϵ_N—the free energy change when a specified amphiphile is put into a micelle with aggregation number N—does not have a minimum value as a function of N, in contrast to the situation for spherical micelles. Away from the ends of the cylinder, the energy of an

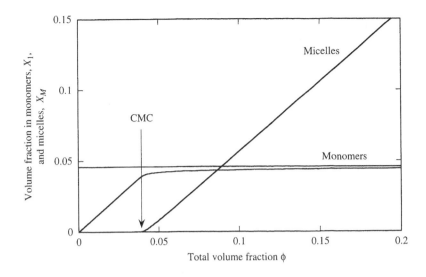

Fig. 9.4 The volume fraction in monomers and micelles as a function of total volume fraction of amphiphile, in the approximation in which all micelles are taken to have the same aggregation number M. Below the CMC, almost all the amphiphile is present as monomers. Above the CMC, as extra amphiphile is added to the solution almost all goes into micelles, with the concentration of monomers rising only very slowly.

amphiphile in the cylinder is independent of the length of the micelle. The only dependence on size comes from the end of the cylinder. Here the molecules must pack with more curvature than they would like, in order to make a spherical end cap to protect the hydrophobic core from contact with water; we suppose that this gives a positive contribution to the energy ΔE_{endcap}. We can write ϵ_N as the sum of a constant term ϵ_∞, representing the energy in the body of the cylinder, and one molecule's share of the end-cap energy, $2\Delta E_{endcap}/N$, which we write as $\alpha k_B T/N$ for compactness. Thus we have for cylinders

$$\epsilon_N = \epsilon_\infty + \frac{\alpha k_B T}{N}. \tag{9.14}$$

If we substitute this into eqn 9.3 we find

$$X_N = N \left[X_1 \exp \alpha \right]^N \exp(-\alpha). \tag{9.15}$$

We see that there is a critical micelle concentration $\phi_c \approx \exp(-\alpha)$; below this volume fraction almost all the amphiphile is present as monomer. To be more precise, we can obtain the total volume fraction by summing the volume fractions present in all micelle sizes, using the identity $\sum_{N=1}^{\infty} N x^N = x/(1-x)^2$:

$$\phi = \sum_N X_N$$
$$= \sum_N N[X_1 \exp(\alpha)]^N \exp(-\alpha)$$
$$= \frac{X_1}{[1 - X_1 \exp(\alpha)]^2}. \tag{9.16}$$

From this, we can find an explicit expression for X_1 in terms of ϕ, which for values of ϕ well above the CMC (i.e. for $\phi \exp(\alpha) \gg 1$) gives approximately

$$X_1 \approx \left(1 - \frac{1}{\sqrt{\phi \exp(\alpha)}} \right) \exp(-\alpha). \tag{9.17}$$

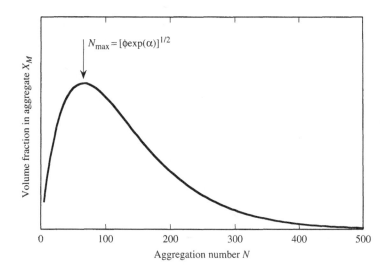

Fig. 9.5 The distribution of aggregate sizes in a system which forms cylindrical micelles, according to eqn 9.18, calculated for $\alpha = 10$ and $\phi = 0.2$.

Thus we find for X_N

$$X_N = n \left(1 - \frac{1}{\sqrt{\phi \exp(\alpha)}} \right)^N \exp(-\alpha). \qquad (9.18)$$

In contrast to the situation for spherical micelles, this is a very broad distribution; the shape is shown in Fig. 9.5. The most probable value of N, N_{max}, is given by

$$N_{\text{max}} = \sqrt{\phi \exp(\alpha)}. \qquad (9.19)$$

A solution of cylindrical surfactants can be thought of as a solution of **living polymers**. Micelles are continuously breaking and reforming; the drive to merge comes from the reduction in end-cap energies that occurs when two micelles merge, while the driving force for long micelles to break into two arises from the extra entropy the system gains by having more micelles.

9.2.6 Bilayers and vesicles

Amphiphiles with a *large* hydrocarbon volume and *small* values of the critical chain length and optimum head-group area will tend to assemble in bilayers. The easiest way of achieving this combination of properties is to have a molecule with *two* hydrocarbon tails attached to each head-group. The most important class of such materials are the phospholipids, such as lecithin, which are of immense biological importance as the major components of cell membranes.

If we have a bilayer-forming amphiphile solution above its CMC, what is the size distribution of fragments of bilayer? We can use an analogous argument to the one we used above to find the size distribution for cylindrical micelles. To find an expression for the free energy change when a specified amphiphile is put into a micelle with aggregation number N, ϵ_N, we argue that the energy of an amphiphile in the disk is independent of its area apart from at the edges. This leads to the expression

$$\epsilon_N = \epsilon_\infty + \frac{\alpha k_{\text{B}} T}{\sqrt{N}}, \qquad (9.20)$$

by an analogous argument to the one we used for cylindrical micelles. Substituting this into eqn 9.3 we find

$$X_N = N[X_1 \exp(\alpha)]^N \exp(-\alpha N^{1/2}). \qquad (9.21)$$

Comparing this with the equation for cylinders, eqn 9.18, we find the factor $\exp(-\alpha)$ replaced by $\exp(-\alpha N^{1/2})$. The factor of $N^{1/2}$ leads to a qualitatively different situation. Rather than there being a distribution of sizes of micelles, as there is for cylinders, in the case of disks we see that the number of large but finite micelles is exponentially small. Above the CMC, any extra amphiphile we add to the system must join an infinite, two-dimensional, sheet-like aggregate.

There is one possibility that we have not considered that could alter this conclusion. If the edges of the sheet can join up on one another, to give a closed surface of bilayer, then the extra energy of the edge may be eliminated. Such a structure is known as a **vesicle**. However, there is an energy cost in forming a vesicle, which arises from the **curvature** that it is necessary to impart to the bilayer. This kind of curvature energy will be discussed in more detail in the next section. This energy cost needs to be set against the energy gain from eliminating the edge, together with a gain in translational entropy that arises because the vesicles are of finite size.

It turns out that, for a system containing only one type of amphiphile, vesicles are stable only at extremely low concentrations. In practice, vesicles can be obtained reasonably easily, for example by breaking up the bilayers in a lamellar phase using ultrasound; these are metastable but reasonably long-lived on experimental timescales. Vesicles can also be formed at equilibrium from **mixtures** of surfactants.

Vesicles are potentially of great importance as a means of **encapsulating** particular molecules. For example, if drug molecules are encapsulated in vesicles they can be delivered to a target part of the body without unwanted interactions; if the vesicle can be made to break up in a controlled way at the target then this provides a very efficient way of delivering the drug.

Vesicles can also be considered to be a simple model of a biological cell, whose contents are separated from the outside world by a bilayer membrane of this kind. This will be discussed in more detail in the next chapter.

9.2.7   The elasticity and fluctuations of membranes

In the last section we introduced the idea of a **bending energy** for a bilayer membrane. Here we introduce a way of quantifying the idea of bending energy, and discuss the fluctuations of a free membrane that occur owing to Brownian motion.

The curvature of a surface at any point can be described by two **principal curvatures** c_1 and c_2. For mathematical details, see for example Safran (1994); here it is sufficient to comment that c_1 and c_2 can be thought of as the coefficients of the quadratic terms in a Taylor expansion of the surface around a point. In Fig. 9.6 this is illustrated by two surfaces. For a locally concave shape, both principal curvatures are positive, while for a saddle shape one curvature is positive and the other is negative. It can be shown that one can write the elastic

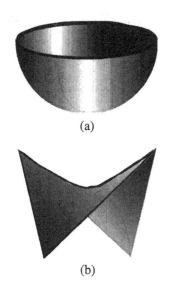

(a)

(b)

Fig. 9.6 Principal curvatures of surfaces. (a) Both principal curvatures are positive; (b) one curvature is positive, the other negative.

energy associated with bending a membrane in the differential form

$$dE_{el} = \left[\frac{k}{2}(c_1 + c_2 - 2c_0)^2 + \bar{k}c_1c_2\right]dA, \tag{9.22}$$

where dE_{el} is the elastic energy needed to distort a differential area of dA membrane to a shape with curvatures c_1 and c_2. Here c_0 is the **spontaneous curvature** of the membrane; this accounts for any tendency of the membrane to bend in one direction or another owing to the local packing constraints of the molecules. For a single layer of surfactants the value of the spontaneous curvature is controlled by the packing considerations described in Section 9.2.3. For a bilayer made of a single component then the spontaneous curvature must be zero; any tendency a single layer has to bend is counteracted by the fact that the other side of the bilayer would have to bend in the unfavourable direction.

The two elastic constants k and $\bar{k}$ are known as the **bending modulus** and the **saddle-splay modulus**. A stable film will always have a positive value of the bending modulus k; this means the membrane is at its lowest energy when the **mean curvature**, $(c_1 + c_2)/2$, is zero. But the saddle-splay modulus may be positive or negative; the system may favour either the shape shown in Fig. 9.6(a), in which the product c_1c_2 (the **Gaussian curvature**) is positive, or saddle shapes such as Fig. 9.6(b), with negative Gaussian curvature, according to the details of the interactions between the molecules.

Equation 9.22 can be used to calculate the energy involved in any distortion of a membrane, and in particular we can use it to evaluate the all-important effect on the shape of the membrane of thermal fluctuations. Before we do this, though, there is one important simplification we can make. This involves the product of curvatures c_1c_2 known as the Gaussian curvature. A remarkable theorem—the **Gauss–Bonnet theorem**—tells us that the integral of the Gaussian curvature over a closed surface is an invariant which depends only on the topology of the surface, not its detailed shape. Specifically, we find

$$\oint dA c_1 c_2 = 4\pi(n_c - n_h), \tag{9.23}$$

where $(n_c - n_h)$ is a topological invariant: the difference between the number of connected components (n_c) and the number of handles of the surface (n_h). Thus for a given shape of surface, as long as we consider only a continuous deformation rather than a change in topology, only the bending modulus comes into play.

So if we consider a bilayer membrane (so that the spontaneous curvature is zero), and consider fluctuations of a large sheet of a kind which involve only distortions of shape without any changes in topology, we find that the bending energy is given simply by

$$E_{el} = \oint \left[\frac{k}{2}(c_1 + c_2)^2\right]dA. \tag{9.24}$$

We can use this expression to evaluate the effect of Brownian motion on the shape of a large, isolated sheet of membrane. Consider a distortion that takes the form of a sinusoidal standing wave of the membrane of wavevector q and amplitude $a(q)$. Using eqn 9.24 to evaluate the bending energy associated with this standing wave we find it to be given by $ka^2(q)q^4/2$ per unit area.

At equilibrium the energy of each such standing wave must be given on average by $k_BT/2$ by the equipartition theorem, so we find for the amplitude

$$\langle |a(q)|^2 \rangle = \frac{k_BT}{Akq^4}, \tag{9.25}$$

where A is the area of the system. At equilibrium the distortions of the membrane caused by Brownian motion will take the form of a superposition of sinusoidal standing wave modes, of random phase and of all wavevectors q, with their average amplitudes being given by eqn 9.25. The membrane has an effective roughness Δz that is found by summing the contributions of all the standing waves:

$$\langle \Delta z^2 \rangle = \int \frac{Aq\,dq}{2\pi} \frac{k_BT}{Akq^4} = \frac{k_BT}{2\pi k} \int \frac{dq}{q^3}. \tag{9.26}$$

Here we have counted all the states in two dimensions with a given magnitude of wavevector $|q|$. The integral in eqn 9.26 **diverges**. The physical significance of this is that if we have a membrane of size L, the effective thickness of the membrane varies as k_BTL^2/k. The thermal undulations of a membrane are so extreme that an isolated membrane, on long length scales, is not a planar object at all but instead is highly **crumpled**. In fact, one can define a **persistence length** ξ_k; on length scales smaller than ξ_k the membrane is locally flat, but on longer length scales the orientation of the membrane wanders randomly. The persistence length can be shown to be given by the expression

$$\xi_k = a \exp\left(\frac{4\pi k}{\alpha k_BT}\right), \tag{9.27}$$

where a is a microscopic length scale, and α is a constant of order unity (Safran 1994).

9.2.8 The phase behaviour of concentrated amphiphile solutions

We have seen that above a certain critical concentration—the CMC—any additional amphiphile that is added to a solution is used to make more micelles. What happens when the concentration is large enough for the micelles to start interacting? A dispersion of spherical micelles is similar in some ways to the dispersions of colloidal particles we considered in Chapter 4. The particles interact via an effective potential; this may have repulsive components arising from excluded volume interactions, screened charge interactions, and steric interactions. The repulsive potential typically leads to the formation of an ordered phase—a colloidal crystal—at high volume fractions. Micelles also interact in the same way, and also have a tendency to form ordered phases at high volume fractions. However, there is a fundamental difference between colloidal particles such as spheres of polymer or silica and micelles: micelles can alter both their **size** and **shape** in response to their interactions with their surroundings. It is this element of mutability that makes the phase behaviour of concentrated amphiphile solutions so rich.

Figure 9.7 shows a (relatively simple) example of the phase behaviour that can be found in amphiphile solutions. This specific example is for

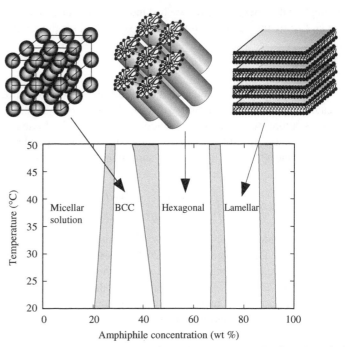

Fig. 9.7 The phase diagram of an amphiphilic copolymer in water, showing schematically the structures of the ordered phase. The material is a short triblock copolymer of ethylene oxide (EO) and propylene oxide (PO), with the structure $(EO)_{37}(PO)_{58}(EO)_{37}$. Data from P. Alexandris, D. Zhou, and A. Khan, *Langmuir*, **12**, 2690 (1996).

a solution of a short block copolymer of ethylene oxide (EO) and propylene oxide (PO); this is a member of a class of materials known commercially as 'Pluronic' or 'Poloxamer' polyglycols, which is frequently used in cosmetics and pharmaceutical applications. The general features of the phase diagram are shared with many other different amphiphilic molecules of all classes.

At relatively low concentration of amphiphile, we find an isotropic micellar solution. This specific molecule has the structure $(EO)_{37}(PO)_{58}(EO)_{37}$, where 'EO' denotes the relatively hydrophilic ethylene oxide units and 'PO' denotes the more hydrophobic propylene oxide units. The molecule thus has two rather bulky hydrophilic units, and so we expect packing geometry considerations to favour the formation of spherical micelles.

As the concentration of amphiphile is increased, repulsive interactions between the micelles begin to be significant. In the case of this material, the repulsion arises both from excluded volume and from the steric effect of the hydrophilic EO brushes. This repulsion leads to a **crystallisation** of the micelles into an ordered structure.

The macroscopic appearance of this phase is a transparent, rather stiff, soft solid. Such materials are often loosely called 'gels', but from the physicist's point of view they are true solids, with long-ranged order that will give rise to sharp diffraction peaks for radiation of the appropriate wavelength. It is this long-ranged order which implies that these materials have a finite (though small) shear modulus, and will only flow when a stress larger than a certain **yield stress** is applied.

The structure of these phases is best probed by diffraction experiments using X-rays or neutrons. These experiments reveal in this case that the structure of the phase is body-centred cubic (BCC), with a cubic cell lattice parameter of 20 nm. This material is a solid, but the repeating unit is two orders of magnitude larger than the atomic sizes which characterise simple solids, and the basic building block of the solid is not an atom or molecule, but an aggregate of many molecules.

As the concentration of amphiphiles increases further, we find a transition to another phase, which diffraction experiments reveal to consist of cylindrical micelles packed into a regular hexagonal array. This is a **columnar liquid crystal** phase, with long-ranged order in two dimensions. Once again, this has the macroscopic appearance of a transparent gel, but the anisotropy of the structure reveals itself in optical birefringence.

Given that the natural curvature arising from molecular packing favours a spherical micelle for this amphiphile, how are we to understand the change in micellar shape? The answer, simply stated, is that cylinders pack more efficiently in space than spheres. The maximum fraction of space that can be occupied by spheres in a BCC structure is 68%, but close-packed cylinders occupy nearly 91%. In this cylindrical phase, then, the system suffers an energy penalty from packing the molecules into micelles with a less than ideal curvature, but this is offset by a decrease in the repulsive energy between micelles that arises from the fact that cylinders pack more efficiently than spheres.

At still higher amphiphile concentrations, there is yet another change in micelle shape, from cylinders to flat bilayers. This is again a liquid crystal phase; it has long-ranged positional order in only one dimension, and thus it is classed as a smectic phase. These materials are transparent; again their anisotropy is reflected in optical birefringence, but in contrast to the cylindrical and cubic phases they can flow under their own weight. The periodicity of the lamellar phase is once again revealed by X-ray diffraction, and for this case is found to be about 10 nm.

We saw in the last section that an isolated membrane would not remember its direction over a distance greater than a certain persistence length ξ_k. How then can we form a smectic phase in which one-dimensional orientational order is maintained over macroscopic distances? In concentrated solutions electrostatic or other direct intermolecular repulsive forces may come into play to keep the layers in order. But even for rather dilute phases order is maintained by another effective force which has an entropic origin, the **Helfrich** force. The origin of this force is to be found in the discussion of the thermal undulations of membranes in the last section. If a membrane is confined not to be able to wander further than a certain distance d from a plane, then a very large number of configurations that otherwise the membrane would have explored during the course of its thermally driven undulations will be forbidden to it, with a consequent loss of entropy. This leads to an effective potential $\Delta F_{\text{Helfrich}}$ that has the form

$$\Delta F_{\text{Helfrich}} \sim \frac{(k_{\text{B}} T)^2}{k d^2}. \tag{9.28}$$

Note that this effective, entropic potential is long-ranged.

9.2.9   Complex phases in surfactant solutions and microemulsions

In the solutions of many amphiphiles, the only phases that appear are drawn from the ones discussed in the last section: micellar solutions, cubic packed micelles, hexagonally packed rods, and lamellar stacks of bilayers. But in some amphiphile solutions more complex phases appear, in which an amphiphile surface is distorted in such a way as to divide space up into two interconnected domains. The resulting structure is called a **bicontinuous network**. Even more complexity is possible in the phase diagrams of ternary mixtures of water, and amphiphile and a hydrophobic liquid, which we will generically term an **oil**. Oil and water do not mix, but in some compositions the presence of amphiphiles can effectively lead to a phase that is stably mixed on a microscopic length scale—a **microemulsion**. One way of thinking about these phases is to imagine them as a micellar solution or mesophase of amphiphiles in water, in which the hydrophobic interior of the micelles, cylinders, or bilayers is swollen by solubilised oil. In the resulting system, water and oil are kept apart by a monolayer of amphiphile.

Complex, bicontinuous phases are probably best thought of as structures that arise from the distortion of a lamellar phase. This distortion is topological in nature, because the resulting phases have quite different connectivities to the lamellar phase. One can imagine introducing pores into a smectic stack of lamellae; the pores would carry an energy penalty arising from bending elasticity, but the system would gain entropy thereby. If the pores proliferate, then the arrangement of the pores may become ordered in order to minimise the resulting bending energy.

One of the remarkable structures that can come about in this way is the structure illustrated in Fig. 9.8. In this structure, the membranes form a regular structure with cubic symmetry. As the structure has long-ranged order, it must

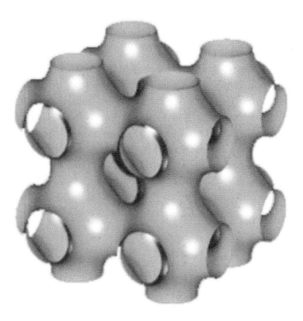

Fig. 9.8 A bicontinuous cubic phase—the cubic P phase.

have a finite shear modulus; such materials have the appearance of being rather weak gels.

One might wonder how such a structure can exist given the large amount of bending energy that surely must arise in it. But the curious mathematical property of such surfaces is that the mean curvature is zero everywhere; at any point the two principal curvatures are equal in magnitude but opposite in sign. The only contribution to the elastic energy comes from the saddle-splay modulus, which may either favour or disfavour such a structure.

What is the effect of fluctuations on such a structure? The presence of thermal energy means that it will be distorted in such a way that the bending energy becomes non-zero. Sometimes this thermal energy can destroy the regularity of this structure, causing the long-ranged order to melt. The result is a random three-dimensional network of tubes known as a **sponge phase**. This structure is sometimes, rather aptly, known as a **plumber's nightmare**.

9.3 Self-assembly in polymers

Metallurgists make new materials by mixing up different elements to make alloys. One might ask if it is possible to do the same with polymers. In some cases one can, and a few important engineering polymers are composed of more than one component—an example is ABS (acrylonitrile butadiene styrene copolymer)—but such materials are not usually simple mixtures of homopolymers. If one takes a pair of common commodity polymers— polyethylene and polystyrene, for example—and mechanically mixes them, the result is that one gets a material with very poor mechanical properties. An examination of the resulting blend would reveal that the two components remained phase separated, and that the interfaces between the two domains were very weak mechanically. In fact, very few pairs of polymers are miscible, and the interfaces between immiscible polymers are almost always extremely weak. This, incidentally, is why it is so difficult to recycle plastics. There are fundamental physical reasons both for the rarity of miscible polymers and for the weakness of polymer/polymer interfaces, and we explore these factors below.

In spite of the difficulties in mixing ordinary homopolymers, it is possible to create chemically heterogeneous polymeric materials by using macromolecules with more complex architectures than simple homopolymers (see Section 5.2.4 for a discussion of different polymer architectures). For example, if one adds to an immiscible blend of two homopolymers A and B an AB block copolymer, this material will segregate to the interface between the two homopolymers. The effect of such segregation is to lower the interfacial tension between the two homopolymers; this is entirely analogous to the way a small-molecule amphile lowers the interfacial energy between oil and water and allows an emulsion to be made. In addition, block copolymers at the interface between droplets of one homopolymer in a matrix of another, immiscible homopolymer suppress the coalescence of the droplets by a mechanism analogous to the steric stabilisation of colloids. Both of these factors tend to lead to a mixture which is still phase separated, but with an average domain size much smaller than it would be in the absence of the copolymer. In addition, the block copolymer may strengthen the interface, resulting in a two-phase material with reasonable mechanical properties.

Block copolymers, and other complex macromolecular architectures in which chemically different lengths of polymer are covalently bound together, provide an even richer range of chemically heterogeneous microstructures. Even if the two chemically distinct parts of a block copolymer chain would be immiscible if they were homopolymers, the fact that the two parts are covalently bound together means that the two components cannot phase separate on a macroscopic scale. Instead, they are constrained to phase-separate only on a microscopic scale, producing well-defined domain structures which often have a high degree of long-ranged order. This microphase separation is closely analogous to the production of complex phases in amphiphile solutions.

9.3.1　Phase separation in polymer mixtures and the polymer/polymer interface

In Chapter 3 we discussed simple models for the free energy of mixing of simple liquids. In the regular solution model, we combined an entropic contribution to the free energy of mixing which derived simply from the translational entropy of the two components with a separate, enthalpic contribution for which we make a mean field approximation. The resulting expression for the free energy of mixing of two monomeric liquids, F_{mono}, eqn 3.6, was

$$\frac{F_{\text{mono}}}{k_B T} = \phi_A \ln \phi_A + \phi_B \ln \phi_B + \chi \phi_A \phi_B. \tag{9.29}$$

How must this expression be modified to deal with polymers? We consider two polymers, each with degree of polymerisation N. Recall that the first two terms in eqn 9.29 represent the entropy of mixing. This arises from the fact that a single site in space is either occupied by a molecule of A with probability of ϕ_A, or by a molecule of B with probability of ϕ_B. If we increase the number of monomer units in the molecule from 1 to N, the entropy of mixing per molecule stays the same, but the energy of mixing must be increased by a factor of N.

The free energy of mixing per polymer molecule $F_{\text{poly}}^{\text{mol}}$ can thus be written

$$\frac{F_{\text{poly}}^{\text{mol}}}{k_B T} = \phi_A \ln \phi_A + \phi_B \ln \phi_B + N \chi \phi_A \phi_B. \tag{9.30}$$

Here the interaction parameter χ represents an energy of interaction per monomer, which we expect to be roughly independent of degree of polymerisation.

It is more usual to write the free energy not per polymer molecule but per monomer unit; writing this as $F_{\text{poly}}^{\text{site}}$ we have

$$\frac{F_{\text{poly}}^{\text{site}}}{k_B T} = \frac{\phi_A}{N} \ln \phi_A + \frac{\phi_B}{N} \ln \phi_B + \chi \phi_A \phi_B. \tag{9.31}$$

This approximation for the free energy of mixing is known in the polymer literature as the **Flory–Huggins** free energy.

To derive the phase diagram for the mixing of two polymers, we now redo the calculations of Section 3.2 using the modified free energy of eqn 9.31 instead of the small molecule regular solution expression of eqn 9.29. In fact, we can see that all our results for small molecules can be transposed to polymers simply by making the substitution $\chi \rightarrow \chi N$. For example, let us consider the critical value of the interaction parameter, χ_c, which separates the situation in which mixtures at all compositions are stable, from the situation in which mixtures at some compositions will phase separate. For small molecules we have

$$\chi > \chi_c = 2 \quad \text{for phase separation}$$
$$\chi < \chi_c = 2 \quad \text{single phase at all compositions,} \quad (9.32)$$

but for polymers we find $\chi_c = 2/N$, that is

$$\chi > \chi_c = \frac{2}{N} \quad \text{for phase separation}$$
$$\chi < \chi_c = \frac{2}{N} \quad \text{single phase at all compositions.} \quad (9.33)$$

As N can be very large for polymers, this means that the size of enthalpically unfavourable interaction that can be tolerated before phase separation happens is very much smaller for polymers than for small molecules. This is why one finds very few pairs of polymers with a large degree of polymerisation that will form single-phase mixtures.

An extreme special case illustrates this point. There is a very small unfavourable energetic interaction between a hydrocarbon and its analogue in which all the hydrogens have been replaced by deuterium. So for a mixture of toluene and deuterium-substituted toluene this interaction corresponds to an interaction parameter $\chi \approx 10^{-4}$; that is to say, there is an energy cost of around $10^{-4} k_B T$ when a molecule of toluene is taken from an environment of toluene and put in an environment of deuterium-substituted toluene. Nonetheless toluene and its deuterium-substituted analogue are miscible in all proportions; this value of χ is substantially less than the critical value of 2. The slightly unfavourable enthalpy of mixing is overwhelmed in the total free energy by the gain in entropy of mixing. But for a polymer such as polystyrene and its deuterium-substituted analogue, while the interaction parameter takes a similarly small value, the critical value of the interaction parameter is a factor of N smaller. It is not difficult to obtain such polymers with degrees of polymerisation in excess of 10^4, and thus we find $\chi > 2/N$, leading to phase separation. This occurs because the entropy of mixing is much less important for large molecules than for small ones.

Most pairs of polymers have much larger values of the interaction parameter χ: values in the range 0.01–0.1 are typical. We can expect such polymer pairs to be phase separated if they are of reasonably large degree of polymerisation. What can we say about such phase-separated mixtures? There are two important points:

(1) the coexisting phases are essentially pure; and
(2) the interface between the coexisting phases is not atomically sharp.

To justify the first statement, one needs to calculate the coexisting compositions from the Flory–Huggins free energy of mixing; what one finds is that for $\chi N \gg 1$ the coexisting composition ϕ_b is well approximated by

$$\phi_b \approx \exp(-\chi N). \quad (9.34)$$

For polymer pairs with reasonably large degrees of polymerisation and a modestly unfavourable interaction parameter χN can easily be considerably larger than unity, leading to coexisting compositions that are to all intents and purposes the pure polymers.

To justify the second statement, we need to recall that a polymer chain is essentially a random walk. An atomically sharp interface would minimise the number of unfavourable contacts between segments belonging to the two different polymers, but this would be at the cost of restricting the number of possible configurations that the chain could adopt, and thereby reducing its entropy. Instead of being atomically sharp the interface must be diffuse to an extent which balances the cost in entropy of having a narrow interface with the increased number of unfavourable interactions between the two polymers in a diffuse interface.

We can estimate the equilibrium width of the interface by a simple argument. Suppose the interface has a width w. Loops of polymer A will protrude into the domain of polymer B (see Fig. 9.9); if the characteristic degree of polymerisation of such loops is N_{loop} then we can relate the loop size to the interfacial width using the fact that the loop is a random walk:

$$w \sim a\sqrt{N_{\text{loop}}}, \tag{9.35}$$

where a is the size of a monomer. Associated with the loop will be an unfavourable energy U_{int} coming from the total number of contacts between A and B segments; there are N_{loop} such contacts so

$$U_{\text{int}} \sim \chi N k_{\text{B}} T. \tag{9.36}$$

But at equilibrium this interaction energy will be of order $k_{\text{B}}T$, so

$$1 \sim \chi N_{\text{loop}}. \tag{9.37}$$

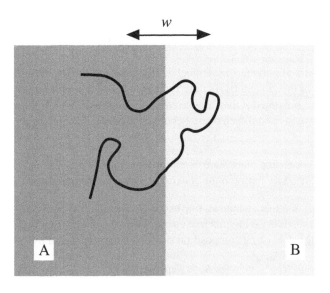

Fig. 9.9 Schematic diagram of the interface between two polymers, A and B. There is a mixed zone at the interface of width w, characterised by loops of polymer A protruding into the domain of B (and vice versa).

If we now use eqn 9.35 to eliminate N_{loop} we find an estimate for the interfacial width w:

$$w \sim \frac{a}{\sqrt{\chi}}. \tag{9.38}$$

Typically this is of order 1–3 nm.

We can also estimate the interfacial energy γ by counting the number of unfavourable contacts per unit area, each of which has an energy $\chi k_B T$:

$$\gamma = \frac{k_B T}{a^2} \sqrt{\chi}. \tag{9.39}$$

This rather crude argument is confirmed by much more sophisticated theories, which yield expressions for the interfacial width and interfacial tensions that are the same, to within numerical factors, as eqns 9.38 and 9.39.

9.3.2 Microphase separation in copolymers

In a block copolymer, two or more chemically distinct lengths of polymer are covalently bound together. If the interaction between the blocks is energetically unfavourable the system will have a tendency to phase-separate. In contrast to the case of mixtures of homopolymers, where the phase-separated domains have a tendency to grow to macroscopic sizes in order to minimise interfacial energy, in a block copolymer the maximum size of the phase-separated domains is constrained by the fact that the chemically different blocks are covalently linked. Thus what results is **microphase separation**.

This is best illustrated in the simple case of a symmetric diblock copolymer, consisting of two chemically different blocks of equal length. Here the balance between the tendency of the two halves of the copolymer to segregate and their constraint leads to a lamellar morphology—a regular alternation of layers of each of the two chemical species (see Fig. 9.10).

The thickness of the lamellae, d, is extremely regular and well defined. What determines its value? Once again, it is given by a balance of free energies. The larger d is, the less area of interface there is per unit volume, and thus the lower the interfacial energy. On the other hand, in order to fit into thick lamellae, the chains have to stretch out to longer than their random walk sizes and this reduces their entropy.

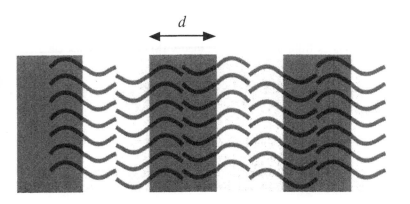

Fig. 9.10 A lamellar microphase-separated morphology in a symmetric block copolymer. The lamellar spacing d is determined by a balance between interfacial energy and the energy of stretching the blocks of the copolymer.

We can estimate the equilibrium lamellar spacing by writing down the total free energy as a sum of the stretching energy F_{el} and the interface contribution F_{int} and finding the value of d that minimises this.

The free energy due to chain stretching per chain is

$$F_{el} \sim k_B T \frac{d^2}{Na^2}, \tag{9.40}$$

so to obtain the stretching energy per unit volume we have a total of $1/Na^3$ chains.

Per unit volume the amount of interface we have is $1/d$, so the total interfacial energy per unit volume is γ/d. The number of chains per unit volume is $1/Na^3$, so the interfacial free energy per chain is

$$F_{int} = \frac{\gamma Na^3}{d}. \tag{9.41}$$

Minimising the total free energy per chain with respect to the lamellar spacing d we find

$$d \sim \left(\frac{\gamma a^5}{k_B T}\right)^{1/3} N^{2/3}. \tag{9.42}$$

If we can assume that the expression we derived for the interfacial tension between immiscible homopolymers, eqn 9.39, also applies here (and more complicated analysis reveals that this is so in the limit $\chi N \gg 1$), then we find

$$d \sim a\chi^{1/6} N^{2/3}. \tag{9.43}$$

9.3.3 Block copolymer phase diagrams

If the two blocks of a diblock are not of equal length, then the lamellar morphology is less favourable, because the shorter blocks end up less stretched than the long ones. This is illustrated in Fig. 9.11. This leads to a tendency for the interface to become curved. Thus as we change the ratio f between the block lengths in block copolymers we will change the equilibrium morphologies.

For strongly immiscible blocks, the morphology changes from lamellar to hexagonally packed cylinders with increasing asymmetry, then it changes to spheres, packed firstly in a BCC array, then in a close-packed array (the difference in energies between HCP and FCC is probably undetectable practically). In addition, more complicated structures have been observed; these are illustrated in Fig. 9.12. It is now possible to calculate the stability of these remarkable phases from first principles theory (Matsen and Bates 1996); the resulting phase diagram is shown in Fig. 9.13.

Further reading

More details on the self-assembly of amphiphiles can be found in Israelachvili (1992), while the theory of interfaces and membranes is compactly presented in Safran (1994). The phase behaviour of polymers and the structure of polymer interfaces are covered in Jones and Richards (1999).

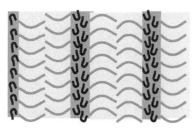

Fig. 9.11 A lamellar morphology (top) is less favoured for an asymmetric diblock copolymer, as it requires the two blocks to be stretched to different degrees. Instead, a morphology with curved interfaces (bottom) is favoured.

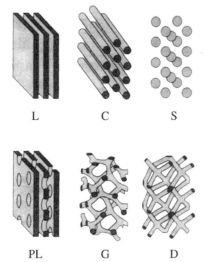

Fig. 9.12 Some block copolymer morphologies. The most commonly observed phases are shown as L, C, and S—lamellae, cylinders, and spheres respectively. PL, G, and D (perforated lamellae, gyroid, and double diamond respectively) are more complex phases that have been more recently identified. Diagram courtesy of M. Matsen.

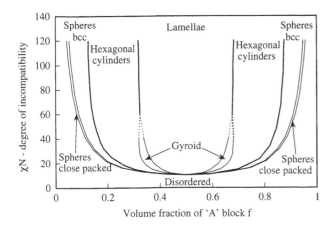

Fig. 9.13 A theoretical prediction of the phase diagram of a diblock copolymer melt, after Matsen and Bates (1996).

Exercises

(9.1) The volume v of a linear hydrocarbon chain with n carbon atoms is given by $v = (27.4 + 26.9n) \times 10^{-3} \text{ nm}^3$, and its critical chain length is $l_c = (0.154 + 0.1265n)$ nm.

a) An amphiphile has an anionic head-group with an optimum head-group area in aqueous solution of $a_0 = 0.65 \text{ nm}^2$.

 i. What shape micelles are formed by amphiphiles with linear hydrocarbon tails with $n = 10$?

 ii. What is the average size and aggregation number of each micelle?

b) In a simple model for ϵ_N, the free energy change when a specified molecule is taken from the bulk and put in an aggregate of N molecules, ϵ_N is written in terms of the head-group area a as $\epsilon_N = \gamma a + k/a$, where k is a constant and γ is an effective interfacial tension.

 i. Derive an expression for the variance in aggregation number of a spherical micelle in terms of γ and the optimum head-group area a_0.

 ii. Evaluate the standard deviation of micelle size for the amphiphile discussed in part (a), assuming that $\gamma = 30 \text{ m Jm}^{-2}$.

(9.2) An amphiphile in a 2.5% solution forms cylindrical micelles. Measurements at lower concentrations of amphiphile indicate a value of the critical micelle concentration of 5×10^{-8}.

a) What is the excess energy associated with an end of the cylinder, in units of $k_B T$?

b) Estimate the most probable aggregation number in a 2.5% solution.

(9.3) a) What is the maximum degree of polymerisation for which blends of polystyrene and PMMA, each having the same degree of polymerisation, are fully miscible?

b) Estimate the width of the interface between high molecular weight polystyrene and PMMA.

c) How many times, on average, is a strand of PMMA in the interfacial region between polystyrene and PMMA, entangled with polystyrene chains?

d) What value of χ would be required between polystyrene and another polymer to ensure that any strand of polymer crossing the interface was entangled with the other polymer species? What would be the interfacial width for this polymer pair?

[Data: The Flory–Huggins interaction parameter between polystyrene and poly(methyl methacrylate) (PMMA) can be taken to have the temperature independent value $\chi = 0.036$; the statistical segment lengths of each of the two polymers may be taken to have the same value $a = 0.67$ nm. The average degree of polymerisation between entanglements is 180.]

(9.4) a) A diblock copolymer of polystyrene and poly (methyl methacrylate) is prepared, in which each block has an identical degree of polymerisation of 3000. Estimate the size of the lamellar spacing.

b) Estimate the degrees of polymerisation for each block of an example of a PS/PMMA block copolymer that will microphase separate into spheres of polystyrene packed in a body centred cubic array in a polymethacrylate matrix.

(9.5) A more sophisticated theory of interfaces between immiscible liquids starts by writing down the excess free energy associated with the interface as a functional of the volume fraction $\phi(z)$, in a form similar to that given in eqn 3.11:

$$F_{int} = A \int \left(g(\phi) + \kappa \left(\frac{d\phi}{dz} \right)^2 \right) dz,$$

where A is the area, κ is the gradient energy coefficient. $g(\phi)$ is given by

$$g(\phi) = f_0(\phi) - f_0(\phi_b),$$

where $f_0(\phi)$ is the free energy per unit volume of a uniform mixture at composition ϕ, ϕ_b is one of the coexisting compositions, and we have assumed that the free energy function is symmetric with respect to composition.

a) Show, using the calculus of variations, that the volume fraction profile $\phi(z)$ which minimises the

interfacial energy obeys

$$2\kappa \frac{d^2\phi}{dz^2} = -\frac{dg}{d\phi},$$

and thus that

$$-g(\phi) + \kappa \left(\frac{d\phi}{dz} \right)^2 = 0.$$

b) Show that the interfacial tension $\gamma = F_{int}/A$ can be written

$$\gamma = \int_{\phi_b^1}^{\phi_b^2} 2 \left[\kappa g(\phi) \right]^{1/2} d\phi,$$

where ϕ_b^1 and ϕ_b^2 are the two coexisting compositions.

c) In a simple theory of an interface between polymers the square gradient coefficient $\kappa = s k_B T/a$, with s a dimensionless constant of order unity and a the statistical step length. Show, using the Flory–Huggins expression for $f_0(\phi)$, that the interfacial tension between two highly immiscible polymers of equal degree of polymerisation N has the functional form of eqn 9.39. Take the volume of a monomer unit to be a^3.

Soft matter in nature

10

10.1 Introduction

Until quite recently almost all the examples of soft matter that were of economic or practical importance were derived from living things. Foodstuffs, for example, are typically systems of colloidal or polymeric character, whose transformations by cooking processes are often best understood in terms of concepts such as those discussed in this book. Materials such as glues and paints were derived from natural products by simple processing techniques carried out on a craft scale; later industrial-scale processes, some still used, extensively modify natural polymers to produce products with less obvious relation to their natural origins. Examples include the modification of cellulose from tree pulp to produce cellophane sheets and rayon fibres, and the use of animal bones and skin to produce gelatin, which apart from its food use is still the main vehicle for photographic emulsions. In the early days of colloid science and polymer science many of the objects of study were essentially biological in origin.

The second half of the twentieth century saw an increasing divergence between polymer and colloid science, on the one hand, and biology on the other. The invention of synthetic polymers and the massive growth of the plastics industry were accompanied by a parallel growth of polymer physics as a subject focused on synthetic rather than natural polymers, while the discovery of the genetic code and the development of protein crystallography led to a new discipline of molecular biology, whose research programme, pursued with spectacular success and intellectual confidence, concerned itself with rather different issues.

Nonetheless, in the picture of life at the molecular level that has emerged from modern biology, we recognise many of the themes of soft matter. Apart from water, the major material components are macromolecular. Systems are dominated by weak, non-covalent interactions, and the fluctuations of Brownian motion are ever present. The formation of the structures necessary for biological function relies on self-assembly processes at both molecular and supramolecular levels. Ultimately, we should be able to hope to understand cell biology in terms of the concepts of soft matter physics. But to do this we have to integrate these concepts with two new principles that are crucial in biology:

1. Unlike the **equilibrium** systems that we are used to dealing with in soft matter physics, living organisms are kept **far from equilibrium** by a constant input of energy.

2. Life **evolves**. In the statistical methods that we often rely on in soft matter physics, we consider an **ensemble** of different realisations of some random system, such as the sequence of monomers along a random copolymer. In biology, we usually need to consider, not the ensemble, but one particular realisation that has evolved to optimise some function. For example, from all the possible random sequences of monomers that make up potential proteins, a few specific sequences have been selected by evolution.

The purpose of this chapter is certainly not to attempt an introduction to biology for physics, or even to summarise the field of biophysics. Rather, it is to highlight some examples from biology in which the ideas explored in previous chapters have been applied.

10.2   The components and structures of life

Life appears to be astonishingly diverse; at first sight there is little in common between a bacterium, a tree and a whale. But underneath the diversity of life's external forms there is a remarkable conservatism in its mechanisms and structures. The basic biochemistry of all life is very similar, and all life is based on the same groups of molecules.

The basic unit of life is the **cell**, an object in which the chemicals needed for metabolism and reproduction are packaged apart from the environment by a membrane based on a lipid bilayer. All life is based on cells; the fundamental distinction at the molecular and cellular level is between **prokaryotes**, consisting of the bacteria, and **eukaryotes**, consisting of plants, animals, fungi and the more complex single-celled organisms known as protocists, such as amoebae and green algae. Prokaryotic cells are relatively unstructured bags of macromolecules contained inside a cell wall. Eukaryotic cells are more structured; they contain membrane-enclosed organelles specialised to carry out certain functions, and they are supported by an internal scaffolding of rigid, linear, macromolecular aggregates—the **cytoskeleton**.

Although the core functions of life are carried out within the cell, the space outside the cell may also be substantially modified for the convenience of the organism. At the simplest level, many bacteria modify their immediate environment by exuding high-molecular-weight, water-soluble polymers to form a slime. In multi-cellular organisms, extra-cellular materials form the major part of animal connective tissues like tendons and bones, and cell walls in plants. These complex, structured materials made from polymers such as proteins and polysaccharides are essential in providing the framework of the organism.

What are the key molecular components of life? Three different classes of polymeric molecules play the crucial roles.

1. **Nucleic acids.** The primary purpose of nucleic acids is to store and transmit the information needed to construct and maintain living organisms. Nucleic acids are **sequenced copolymers** of four different monomer units—the nucleotides. The **genetic code** translates the sequence of monomers in DNA to a sequence of amino acids in proteins. One type

of nucleic acid—DNA (deoxyribonucleic acid)—provides an archive of the information needed to construct an organism, while another type—RNA (ribonucleic acid)—is involved in the transcription of this information and the synthesis of proteins.

2. **Proteins**. Proteins are sequenced copolymers of amino acids, whose sequence is coded by the sequence of nucleotides in DNA. Proteins, together with membranes, provide most of the hardware of cells, with three key functions. **Enzymes** are proteins that function as catalysts, facilitating all the chemical reactions that the cell needs to process energy and synthesise its component molecules (including proteins themselves). Proteins also serve as important **structural elements**, both within the cell and in the extra-cellular parts of tissues such as bones and tendons. Finally, combining their catalytic and structural roles they form the components of **molecular machines** which, for example, can convert chemical energy to mechanical energy both to transport materials around a cell and to give an organism motility.

3. **Polysaccharides**. These are polymers of various sugars. Unlike nucleic acids or proteins their monomer units do not have a regular sequence; instead they are better thought of as random copolymers. Even if they are made from a pure monomer such as glucose there remains a considerable degree of quenched disorder due to randomness in the stereochemistry and random branching. They are used as stores of energy (e.g. starch in plants), and as structural elements (e.g. the cellulose of plant cells or the chitin of insect exoskeletons). Polymers in which polysaccharide side chains are grafted onto a protein backbone (glycoproteins), or in which the polysaccharide forms the main chain to which proteins are grafted (peptidoglycans), are important, the latter as components of bacterial cell walls, and the former in the extra-cellular matrix, and as mucus.

In addition to these polymeric components, life depends on membranes self-assembled from a variety of amphiphilic molecules. The most important of these are **phospholipids**; each of these molecules has two hydrocarbon chains attached to a polar head-group. These molecules, with their large hydrocarbon volume and short critical chain length, are ideally suited to produce stable bilayer structures. The bilayers are then substantially modified by the inclusion of membrane proteins; some of these extend right through the bilayer, while others are anchored to the membrane.

10.3 Nucleic acids

DNA is possibly the most famous biological macromolecule of all. The elucidation of its structure—the **double helix**—by Watson, Crick, Franklin, and Wilkins also revealed the essence of its function as the basis of heredity.

The backbone of a DNA molecule consists of alternating units of phosphate and the sugar deoxyribose (see Fig. 10.1). Attached to each sugar is one of four possible **bases**. These four bases are thymine, cytosine, adenine and guanine (commonly abbreviated as T, C, A, and G). It is the sequence of these bases along the backbone that contains the information of the genetic code—the genome—which allows a new organism to be constructed. More specifically, the order

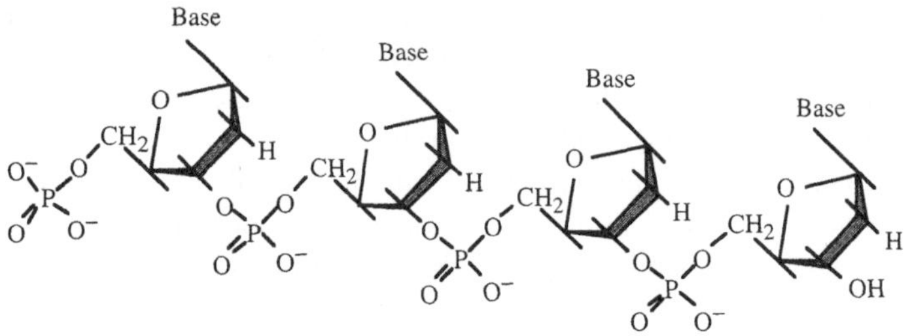

Fig. 10.1 The backbone of DNA, showing the alternation of phosphate groups and the sugar deoxyribose.

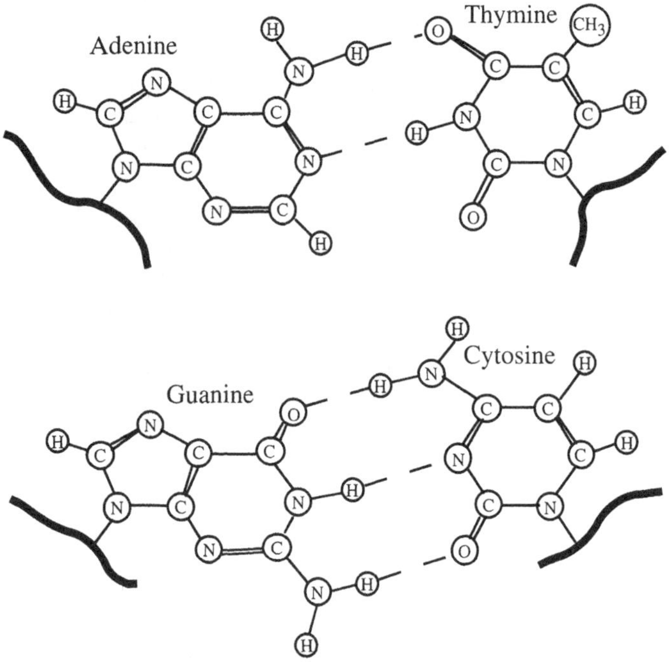

Fig. 10.2 The four bases of DNA, shown as the complementary pairs. The thick lines represent the attachment to the sugar–phosphate backbone of the strand, while the dashed lines represent the hydrogen bonds which hold the complementary pairs together.

of bases encodes the amino acid sequences of all the types of protein that the organism synthesises.

The bases are illustrated in Fig. 10.2. This figure also illustrates the key feature that underlies the functionality of nucleic acids: there are strong and specific hydrogen bond interactions between the pairs of bases adenine and thymine, and cytosine and guanine. Thus these species form **complementary base pairs**. In the native form of DNA two strands with a complementary sequence of nucleic acids come together to form a double helix (Fig. 10.3), in which each base forms a pair with its complement.

The double helix is stabilised both by the hydrogen bonding between the base pairs and by a favourable interaction between adjacent bases along the strand (base stacking). A smaller destabilising influence arises from interactions between the charged phosphate groups; in an aqueous solution containing other ions these charge interactions are screened, as described in Chapter 4, so we can expect to be able to tune the size of the interaction by changing the ionic strength of the solution. The precise free energy change when a complementary base pairing is made also depends both on which base pair is involved and what

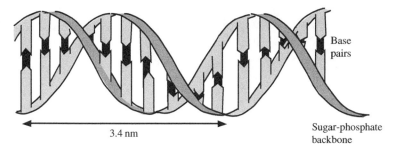

Base pairs

3.4 nm

Sugar-phosphate backbone

Fig. 10.3 A schematic of the double helix of DNA.

the neighbouring species are, but a rough estimate is 1.6 kilocalories per mole at room temperature ($6.7\,\mathrm{kJ\,mol^{-1}}$). This corresponds to about $3\,k_B T$ per base pair. The comparability of this pairing energy with thermal energies suggests that we might observe an experimental transition between the helix form and a denatured form in which single strands exist in random coil configurations. This is indeed observed; typically a rather sharp transition from a double helix to random coils—'helix melting'—is observed in the range 50–90 °C.

The transition between single-stranded DNA in a random coil form and the double helix is highly reminiscent of the helix–coil transitions for single chains we discussed in Chapter 7, and can be treated by a very similar theory. One difference arises from the fact that the helix has two strands rather than one. This means that, if a section of coil is formed in the middle of a stretch of helix, the two strands have to come together at each end of the coil section to form a closed loop; this results in a lower entropy than a section of random coil formed by a single chain. The effect of this is to make the helix–coil transition sharper for a double-stranded helix such as DNA than for a single-stranded helix.

A DNA double helix is necessarily rather a stiff structure, but the sheer length of DNA molecules in living systems means that it is wrong to think of the molecule as a rigid rod; even the very small degree to which the molecule can bend is big enough that on large length scales the polymer does form a random coil structure. This idea can be made precise in terms of a model known as the 'worm-like chain'. Without going into detail, we consider the molecule to be a stiff but not rigid rod with a certain bending modulus. Thermal fluctuations of the molecule mean that there is a loss of correlation of direction over a certain distance known as the **persistence length** l_p. On length scales smaller than l_p, the molecule is effectively rigid, but for larger length scales it behaves like a random walk with an effective step length that is very much larger than the actual distance between monomer units (recall the discussion of effective step lengths in Section 5.3.2 in the context of the jointed chain model with fixed angles of rotation). The theory of the worm-like chain model shows that the end-to-end distance is given by

$$\langle \mathbf{r}^2 \rangle = l_p L. \tag{10.1}$$

Here L is the total contour length along the chain, which is written in terms of the number of monomer units N and the size of a monomer a as $L = Na$. By comparison with eqn 5.9 we can see that the statistical segment length b is given by $b = \sqrt{(l_p a)}$.

For DNA, the persistence length depends on the ionic strength of the medium, but in a typical situation we find values of the order $l_p \approx 60\,\mathrm{nm}$.

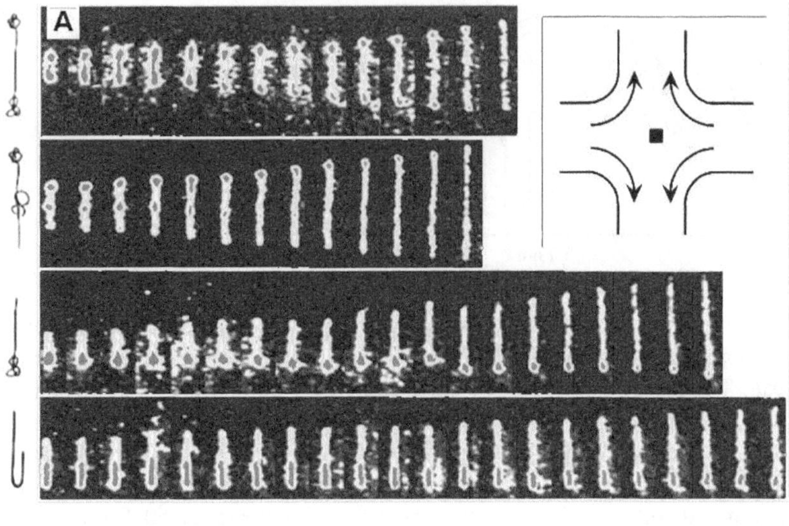

Fig. 10.4 Stretching single molecules of DNA. A solution of DNA molecules from a bacteriophage is fluorescently labelled and is put into the flow cell sketched top right. The flow field causes the molecules to stretch out; individual molecules are imaged using fluorescent microscopy. Reprinted with permission from Perkins *et al.* (1997). Copyright 1997 American Association for the Advancement of Science.

This is very large: the distance between monomer units is around 0.34 nm. The DNA of a bacterium such as *E. coli* has 4.7×10^6 monomer units, giving a total contour length of 160 micrometres and an end-to-end distance as an ideal random walk of 230 nm. Each human cell contains 24 distinct DNA molecules (one per chromosome); the largest of these has 280×10^6 monomer units, giving a total contour length of 95 mm, and an end-to-end distance as an ideal random walk of 75 μm.

The huge size (by normal molecular standards) of the DNA molecule has a number of interesting consequences, of which we highlight two.

1. Single DNA molecules are easily visible in a light microscope, if they are labelled with fluorescent compounds. This is the basis of some strikingly elegant demonstrations of polymer dynamics (see Fig. 10.4).

2. Whereas the relatively short DNA molecule of a bacterium could conceivably be accommodated within the dimensions of a cell, the DNA of eukaryotic organisms such as humans is far too large to be accommodated in a cell in its random coil configuration, and must instead be packed, together with a number of protein components, in a complex, hierarchically structured, bundle—the **chromosome**.

The details of the way in which DNA is packed in living organisms are too intricate for us to consider here. However, there is one more general feature of the DNA molecule which should be mentioned. In understanding the flexibility of the DNA double helix, we needed to account for the fact that the molecule, although fairly rigid, could still bend appreciably. We could also imagine **twisting** the DNA double helix, this twisting being resisted by a certain torsional stiffness. If one constrains the molecule to have a certain degree of twist, it may respond by forming a **supercoil**. This is most easily visualised by playing with something like an electrical cord. As one twists the cord, it can relieve its elastic twisting energy at the expense of some bending energy by forming a supercoil. If one were now to join together the two twisted ends, one would have a loop with one permanent supercoil. Such closed loops do occur quite often in natural DNA.

10.4 Proteins

Proteins constitute the physical realisation of the information stored in the sequence of bases in the DNA that makes up an organism's genome. The sequence of these bases in a single **gene**—a stretch of the DNA molecule that codes for an individual protein—specifies a completely determined sequence of **amino acids** that are polymerised to express the protein. The function of a protein depends on its shape in **three dimensions**, whereas its sequence represents strictly linear, one-dimensional information. The process by which this one-dimensional code determines a unique structure in three dimensions is another example of self-assembly; this process, **protein folding**, represents one of the most important outstanding problems at the interface between soft matter physics and biology.

We can summarise the simplest view of biology by saying that the genetic code is realised in the physical form of proteins, whose three-dimensional structure controls all aspects of the organism. In reality, there is considerable extra complexity:

- Genes are not all read at the same time and at the same rate. Elaborate external mechanisms exist to control the expression of particular genes.
- The self-assembly of the proteins and of structures formed from proteins may need to be assisted by mechanisms involving other proteins. For example, so-called **chaperones** are often necessary to permit correct folding, and templates may be needed to build up more complex structures from many protein molecules.
- Following synthesis and folding protein chains may be further chemically processed inside or outside the cell. Some proteins, for example the collagen that is an important structural component in bones and tendons, are synthesised in the form of a longer, soluble precursor chain which has to be cut to yield the strong, insoluble final material, while in many other cases reactions with polysaccharide chains occur after synthesis to yield graft copolymers of proteins and polysaccharides (glycoproteins and peptidoglycans).

These and other factors should warn one that the simple linear chain of causality implied by the progression genes → protein sequence → protein structure → function conceals many complex interactions and feedbacks.

10.4.1 Primary, secondary, and tertiary structure of proteins

The basic monomer unit of a protein is an amino acid, which has the generic form shown in Fig. 10.5. Polymerisation of amino acids leads to a backbone of carbon atoms linked by peptide bonds, the same generic structure as synthetic polymers in the nylon family. What gives proteins their variety and special properties are the different possible side groups. In nature 20 different amino acids are used to make proteins (many other different ones can be made synthetically). These may be classified as follows:

- Hydrophobic.
- Hydrophilic, uncharged.

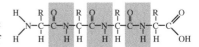

Fig. 10.5 (Top) The structure of an amino acid. R can be one of 20 different side groups. (Bottom) The backbone of a protein, showing the peptide bond linking successive amino acid residues (shaded).

- Hydrophilic, acidic.
- Hydrophilic, basic.

The linear sequence of amino acids is known as the **primary structure** of the protein.

The combination of hydrophilic and hydrophobic units along the length of the chain makes a protein in solution in water a particular kind of amphiphile. Just like the amphiphilic molecules described in Chapter 8, there will be a tendency for the molecules to form structures in which the number of contacts between hydrophobic groups and water is minimised, and the number of contacts between hydrophilic groups and water is maximised.

A folded protein molecule can thus be thought of as a particular kind of single-molecule micelle. An important difference between a folded protein and an ordinary micelle, formed say from soap molecules, is that, whereas there are many possible arrangements and conformations of the soap molecules in the micelle, in the folded protein there is a single, well-defined structure—the **native state**. We shall consider some of the underlying reasons for this fundamental distinction in the next section.

The core of the folded protein is essentially close packed in density. Within this core sections of the chain often organise themselves in certain recurring arrangements; these common structural elements are known as **secondary structure**. The two main classes are:

- α-helix. In this structure, which we already have introduced in Section 7.7.2, the polypeptide main chain forms a helix which is stabilised by hydrogen bonds between the C=O group on the ith monomer and the N–H group on the $(i + 4)$th monomer. The result is a rather rigid, rod-like structural element.
- β-sheet. Here successive parts of the polypeptide chain form sheets in which the neighbouring strands are joined by hydrogen bonds (Fig. 10.6). The result is a structure analogous to one plane of a chain-folded crystal of the kind introduced in Chapter 9.

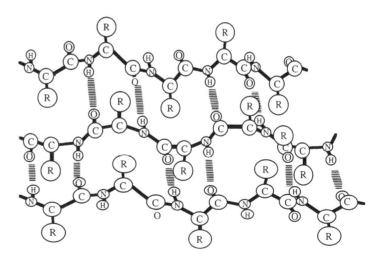

Fig. 10.6 The structure of a β-sheet; three strands of polypeptide run in a folded plane, with neighbouring strands anti-parallel. Hydrogen bonds stabilise the structure by linking neighbouring peptide bonds.

One point that is worth stressing is that both these structures arise from the packing of the main polypeptide chain, independent of the side groups. The chemical nature of the side groups and the interactions between them will only modify the stability of these structures.

The **tertiary structure** of the protein comprises the full three-dimensional arrangement of the chain and all its side groups. Once again, it should be stressed that unlike a micelle, the position of each monomer and its side group is uniquely specified in this tertiary structure, although the importance of Brownian motion does imply that the structure should not be thought of as absolutely rigid.

Tertiary structures of many proteins have now been experimentally determined; the classical technique for doing this is X-ray crystallography. The advent of powerful synchrotron sources combined with sophisticated data analysis software has made this quick and accurate, though the technique still relies on being able to produce protein crystals of a reasonable size. Smaller proteins may now be characterised in solution by NMR. An example of a protein structure is shown in Fig. 10.7.

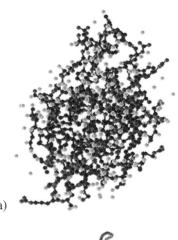

(a)

(b)

Fig. 10.7 The structure of a globular protein (hen egg lysozyme) as determined by X-ray diffraction. In (a) the structure is represented by a ball-and-stick model, while in (b) the same information is conveyed in a more schematic way, showing strands involved in β-sheets as broad arrows and strands involved in α-helices as helical sections. Figures generated from coordinates in the protein database, structure 1LYZ, Diamond *et al.* (1975).

10.4.2 Protein folding

If a solution of a globular protein such as the one shown in Fig. 10.7 is heated up, there will usually be found a well-defined temperature at which the protein ceases to be in its compact, native state, and instead takes up a much more open, random conformation. This loss of structure is associated with a loss of biochemical activity; an enzyme, for example, whose catalytic activity depends on the precise shape adopted by the molecule in its native state, will cease to work. This process is known as **denaturation**. If a solution of protein is unfolded in this way, and the concentration of protein is low enough, and the solution of unfolded protein is subsequently cooled down again, it is sometimes found that the protein **refolds** into its native state again, recovering its biological activity. This suggests that there is a well-defined transition between the native state and an unfolded state: the **unfolding** transition.

The character of this transition can be further elucidated by carrying out careful thermodynamic measurements. These confirm that the transition is reversible, and it has associated with it a certain latent heat. The transition is a true first-order thermodynamic phase transition, and is analogous to melting. The fact that the transition takes place in a small system—a single protein molecule—does mean that rather than being infinitely sharp the transition takes place over a finite range of temperatures.

It is tempting to compare protein folding with other, simpler, transitions that can occur in single-polymer chains. We have met two such transitions earlier, the helix–coil transition and the collapse of a single chain in a poor solvent.

We have seen that α-helices are a significant part of the secondary structure of many proteins, and the classical systems for observing the helix–coil transition are synthetic polypeptides such as poly(benzyl-l-glutamate). Despite these affinities the helix–coil transition is fundamentally different in character to protein folding. The theory outlined in Section 7.7.2 makes it clear that the helix–coil transition can never be a true thermodynamic phase transition; the finite width of the transition persists even in the limit of infinitely long chains.

This contrasts with the protein-folding transition, which is a true phase transition that only has a finite width because the chains are of a finite size.

The coil–globule transition that occurs when a chain collapses as solvent conditions are changed from good to poor was introduced in Section 5.3.3. There it was stated that in the limit of long chains this was a true phase transition, which could be either first or second order according to circumstances. In the case of a protein, as the temperature is increased we expect the hydrophobic effect to become less strong, so as the protein solution is heated the balance of hydrophobic and hydrophilic interactions is changed from an overall bad solvent condition at low temperature to good solvent conditions at high temperature. However, there are still vital aspects of the protein-folding transition that are not captured by the analogy with the coil–globule transition. Most crucially, in the globule state of a simple homopolymer there are many different conformations with roughly the same free energy, and one can consider the globule to be a liquid in which these different conformations are continually being sampled. In a protein in its globular form, however, there is only one conformation—the native state.

A fully realistic model for protein folding would calculate the free energy of each possible configuration of a realistic representation of a protein chain and all the interactions between its amino acid units. Even with the most powerful computers this is not possible, nor is it likely to be for some time. What is more, there are general questions that would not be answered by doing this kind of calculation for specific protein sequences. In particular, for a full understanding of protein folding one would wish to know not only whether and how a particular protein sequence folds, but how many of all the possible sequences of amino acids have a well-defined native state. These issues can begin to be addressed in much simpler models of random heteropolymers.

A computationally tractable model that has proved very important in getting an understanding of protein folding represents the molecule by a discrete sequence of monomers that are confined to a lattice. We represent the interactions by a contact interaction between neighbouring monomers on the lattice. In the simplest form of the model we could simply have two types of monomer—hydrophobic and hydrophilic—with a lower energy of interaction between like monomers than between unlike monomers. For relatively short chains, one can then simply enumerate all the possible configurations of the chain and calculate the free energy of that configuration. This suggests a very important concept—that of the **energy landscape** (see for example Dinner *et al.* (2000) for a review). This is a conceptual space in which, for any given sequence of monomers, we plot the free energy as a function of coordinates which represent the different possible configurations. This space is highly multi-dimensional, but we can get some idea of its important features in a two-dimensional representation; here we plot along the x-axis some measure of configurations such that similar configurations are closer together than more unrelated configurations.

Examples of possible energy landscapes are sketched in Fig. 10.8. In Fig. 10.8(a), we have rather a smooth landscape with a well-defined minimum in energy. This corresponds to a situation in which similar configurations tend to have similar energies, and in which there are groups of rather similar configurations that have a minimum energy. This kind of landscape would

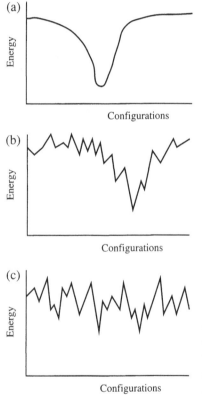

Fig. 10.8 Schematic diagrams of energy landscapes for heteropolymers. (a) represents a smooth landscape with a well-defined minimum; here closely related configurations have similar energies. (b) represents a rough landscape with a well-defined minimum; this would be typical of a protein-like heteropolymer that has a single native state. (c) represents a rough landscape with no well-defined minimum; here there would be no well-defined native state.

be found for a homopolymer in a poor solvent; the high-energy configurations would be the expanded, random coil ones, while for configurations of higher polymer densities one expects a smooth decrease in the energy.

Figures 10.8(b) and 10.8(c) show much rougher energy landscapes. Here, quite closely related configurations can have quite different energies. It turns out that this situation is much more typical for heteropolymers with random sequences of monomers. Why should this be? The reason is that random heteropolymers, in which there are widely varying interaction energies between different monomer types, have a characteristic known in theoretical physics as **frustration**. In order to achieve lowest energy one would need to pack similar monomers close together, but because all the monomers are joined firmly, the system has too many constraints for them all to be satisfied simultaneously. In this situation a relatively small change in configuration can produce quite a large change in energy.

An obvious characteristic of rough energy landscapes is that they need not have a well-defined minimum in energy, unlike the smooth landscape of Fig. 10.8(a). The landscape of Fig. 10.8(b) does have such a minimum; this heteropolymer would be protein-like in the sense that it has a well-defined ground state, but in Fig. 10.8(c) there are many states widely separated in configuration space with similar energies; this molecule would have no single native state. What is found from studying heteropolymers on a lattice is that of the many possible random sequences of monomers, by far the majority have energy landscapes of the form shown in Fig. 10.8(c); heteropolymers that do have a single native state are in a tiny minority.

How is it, then, that proteins have a well-defined native state, when the overwhelming majority of random sequences of amino acids would not have such a native state? The answer is that protein sequences are not random, they have **evolved** specifically to have this property. The driving force for this molecular evolution must be the optimisation of properties such as catalysis.

The energy landscape picture also allows us to rationalise the **kinetics** of protein folding. Given the very large number of possible configurations in three-dimensional space of a given protein sequence, it was always a puzzle that a protein could search enough configurations to find the native state in the timescale that we know experimentally proteins take to fold (this is usually of the order of milliseconds). This puzzle is known in the biophysical literature as the Levinthal paradox. The solution to the paradox is that the folding process does not represent a completely random search through all possible configurations; rather the search is biased towards states of successively lower energy. In the energy landscape of a foldable protein, a **funnel** of states of successively lower energy guides the search of the protein towards the native state. This is illustrated in Fig. 10.9, which shows energies and configurations calculated for the folding process of a lattice heteropolymer. The system starts at high temperature in one of many possible open configurations all of which have equivalently high energy. In this state the system has a high entropy, as there are many configurations with the same energy. When the temperature is lowered the system explores increasingly compact states with lower energies; at each stage small changes in configuration smoothly lower the energy until the native state is reached.

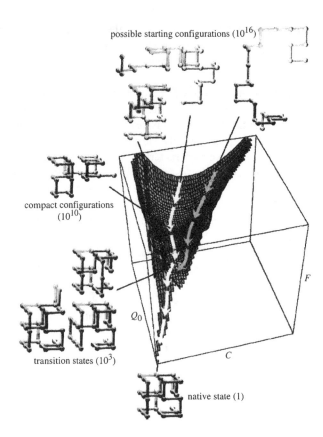

possible starting configurations (10^{16})

compact configurations (10^{10})

F

Q_0

C

transition states (10^3)

native state (1)

Fig. 10.9 The folding funnel illustrated by a lattice simulation of heteropolymer collapse. Reproduced, with permission, from Dinner *et al.* (2000). Copyright (2000) Elsevier Science Ltd.

10.4.3   Interactions between proteins: misfolding, aggregation, and crystallisation

Experimentally, refolding of an unfolded protein will only usually occur if the protein is present at rather low dilution; protein folding is a single-chain process. If one heats up a protein solution at higher concentrations—for example, at the concentration of an egg-white—interactions between protein molecules become important. A heated egg-white sets owing to the development of irreversible associations between its component protein molecules. A simple analogy is with ordinary homopolymers in solution; at low concentrations in a bad solvent interactions between segments within a single molecule cause the chain to collapse, but at higher concentrations interactions between segments belonging to different chains become more important and lead to phase separation. The range of different kinds of interactions between protein molecules is of course much richer than this.

We can distinguish between two different categories of interaction between proteins:

1. Interactions between proteins whose native state remains largely unperturbed. These include **protein crystallisation** and the self-assembly of globular proteins to form filaments, sheets, and tubes.
2. Interactions leading to states in which the protein molecules adopt a conformation which is completely different to their native state. Such states include **heat-set gels** and **amyloid fibrils**.

Protein crystallisation

A solution of globular proteins in conditions far from the unfolding transition has much in common with the monodisperse colloidal dispersions discussed in Section 4.4.1. The protein molecules are in the form of relatively undeformable, roughly spherical objects that interact with each other via charged groups present on the surface as well as through the excluded volume interaction. As such they form crystals in which the protein molecules, together with water, pack together in regular arrays. Protein crystallisation as a phenomenon is of negligible importance in living organisms, though it is of central importance to structural biology, as it is only by growing high-quality crystals of reasonable size that the three-dimensional structure of proteins can be determined at atomic resolution using X-ray diffraction. Current techniques for growing crystals combine a considerable degree of craft skill with sophisticated automation. One of the approaches which has been widely publicised as a means of growing better protein crystals is the use of microgravity conditions in a space station. It will probably prove more effective (as well as much cheaper) to try and use the fundamental understanding gained in the context of colloidal crystallisation (as discussed in Chapter 4) to find ways of modifying the forces acting between protein molecules to optimise the crystallisation process.

Self-assembly of proteins into filaments, sheets, and tubes

Many structures in the cell are formed by the self-assembly of globular proteins into larger units. Important examples include **actin filaments**. These are formed by the intermolecular association of a protein consisting of 375 amino acid units into a fibre with a diameter of around 6 nm; this fibre resembles a twisted pair of wires with a helix pitch of 37 nm. **Microtubules** are built up from a pair of proteins which associate first into dimers, and then into a hollow tube with a diameter of about 25 nm. Actin filaments and microtubules form part of the **cytoskeleton** of eukaryotic cells; in addition microtubules are the main structural component of the cilia and flagella that are found at the surface of many animal cells and that are the mechanism for many single-celled animals and sperm cells to swim.

The starting point for understanding self-assembly is the discussion in Section 9.2 of the self-assembly of amphiphiles. There is some interaction between the units which leads to a lowering of energy when a new unit joins a growing aggregate, and against this must be offset the free energy cost associated with a loss of the translational entropy of the bound unit. A simple statistical mechanical analysis leads to the idea of a **critical aggregation concentration**, analogous to the critical micelle concentration (CMC) of an amphiphile system.

Some remarkably complex biological structures are self-assembling in the strictest sense; that is, if one assembles in a test tube the component molecules the structure will spontaneously emerge. The best known examples of such structures are some viruses, in which proteins assemble to form a complete coat enclosing the genetic material of the virus. The assembly of both actin and tubulin has a further level of complexity, however. In both cases the aggregation process is coupled to the hydrolysis of an energy-containing molecule. This means that aggregation can be switched on and off, and indeed that it can be used to do mechanical work.

The way this works is best illustrated with actin. The actin molecule binds a molecule of ATP (adenosine triphosphate) which can be hydrolysed to ADP (adenosine diphosphate), with the release of some chemical energy. Effectively the binding energy of the actin depends on whether it is associated with ADP or ATP; the bonds are weaker in the form associated with ADP than in the ATP-associated form. If actin is present in the unaggregated form in a test tube, and ATP added, this will lead to the rapid polymerisation of the actin and the formation of filaments. ATP is unstable, however, and will decompose, once again driving the actin filaments to dissociate. In this way a cell can control the number and density of actin filaments, and thus the viscoelastic properties of the cytoskeleton can be precisely tuned.

10.4.4   Protein misfolding, gelation, and amyloidogenesis

As we discussed above, it is believed that a single-protein molecule has a single well-defined conformation of lowest free energy—the native state. But it is not at all obvious that the lowest free energy state of a solution of protein molecules at finite concentration is an assembly of individual proteins each independently folded. In circumstances in which proteins are allowed to unfold at finite concentrations, the resulting state is an ill-defined condition known to biochemists as 'denatured'. As the name suggests, it is clear that in this state the proteins are not in their native state, and biological function, such as enzyme activity, that depends on the correct folding of the proteins, will be absent.

These denatured states have not, until rather recently, received much attention from biologists—after all, they represent an experiment that has gone wrong, yielding an unusable product. For example, in biotechnology, when bacteria such as *E. coli* are genetically engineered to express a particular protein, the result is sometimes an insoluble mass of the denatured protein—known as an inclusion body—from which it is impossible to extract a useful product.

Some study of denatured proteins has been made in the context of food science. For example, the whey that remains when cheese has been made from milk consists largely of a solution of globular proteins such as lactoglobulin, together with the sugar lactose and salts. Whey can be converted to a type of soft cheese by exposing it to high temperatures and acid conditions (the Italian cheese Ricotta was originally made thus). This kind of product (known more scientifically as a **heat-set-gel**) is best thought of as a colloidal gel of the kind introduced in Section 4.4, in which particles stick together when they collide with a strong interaction that does not permit rearrangement. Scattering experiments confirm that the aggregates that make up the gel have a fractal character. What are the units that come together to form these large-scale aggregates? It is clear that they are not the individual globular proteins themselves; instead each of the aggregating units consists of a number of protein molecules that have denatured and formed a strongly associated cluster. The clusters, although denatured, are not without structure; it is found that during this denaturing the relative amount of protein chain folded in β-sheets is substantially increased.

The theme of association between protein molecules via the formation of intermolecular β-sheet has recently received new relevance through the recognition of the importance of **amyloid** diseases. These form an important

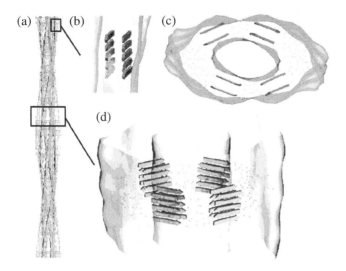

Fig. 10.10 The structure of an amyloid fibril as reconstructed from electron microscopy studies. Reproduced with permission from Jimenez *et al.* (1999). Copyright (2000) Oxford University Press.

and usually incurable class of degenerative diseases, including Alzheimer's disease, type II diabetes, bovine spongiform encephalopathy, and its human analogue Creutzfeldt–Jacob disease. What is common to all of them is that increasing amounts of protein misfold to form insoluble, fibrillar aggregates known as **amyloids**. In each disease a particular protein is involved in the formation of fibrils, and there appears to be little in common between the different proteins that are associated with amyloid diseases. Moreover, other proteins that are not implicated in known diseases also seem to be able to form fibrils very similar to amyloids in the test tube. Structural investigations reveal strong similarities in the structure of different amyloids; in each case the fibrils are stabilised by the extensive, quasi-regular formation of intermolecular β-sheet, with the hydrogen bonds involved being parallel to the direction of the fibril. An example of a postulated structure is shown in Fig. 10.10.

It is a possibility, then, that the amyloid fibril represents a generic state of order in proteins. What is not yet clear is the thermodynamic status of this structure. Is it like the self-assembled structures discussed in Chapter 9, an equilibrium state, or is it like the chain-folded lamellae discussed in Chapter 8, under kinetic control? These questions must await further study.

10.5 Polysaccharides

Proteins and nucleic acids differ from synthetic polymers by the precision of their architecture and in particular their sequence of monomers. The third great class of biopolymers—polysaccharides—lack this precision and in this sense can be considered to be much more directly comparable to synthetic analogues. Nonetheless, despite the lack of precise sequence control, differences in architecture mean that polymers made from the same monomer can have quite different physical properties and biological uses.

If only in volume terms, those polysaccharides which are simple polymers of the sugar glucose are by far the most important. The two polymers which comprise **starch**—**amylose** and **amylopectin**—are used by plants to store energy, while **glycogen** is an analogous storage polymer in animals. **Cellulose** is the major structural polymer in plants. Although the chemical composition of these four polymers is identical, their architecture is different. Both amylose and cellulose are linear, but the stereochemistry of the linkage of the glucose units along the chain is different, and this leads to a very different supramolecular organisation. In cellulose, hydrogen bonds within the chain lead to a stable, ribbon-like structure being formed. Further hydrogen bonds between the ribbons lead to the formation of highly ordered crystalline aggregates—microfibrils. In this form cellulose is highly insoluble and resistant to enzymatic attack. Amylose, on the other hand, forms a much more flexible molecule which is soluble in hot water and much more amenable to degradation by enzymes. Amylopectin and glycogen have the same main chain stereochemistry as amylose, but both have a certain amount of branching. Amylose and amylopectin are laid down by plants in the form of starch granules, which have a rather complicated hierarchical structure. It is the dissolution of these starch granules to release free chains to solution that is the important stage in cooking starchy foods such as potatoes and grain.

More soluble polysaccharides are formed by the polymerisation of sugars with charged groups. In plants, such polymers glue together the cellulose microfibrils to make tough composite structures; many of them are collectively known as **gums**, examples being the pectins that are familiar to jam-makers and the algins that are important components of seaweed. In animals, the extracellular space in connective tissue such as tendons is filled with a swollen gel of charged polysaccharides such as **hyaluronic acid**. The osmotic pressure of these gels makes them very efficient at resisting compression; the production of hyaluronic acid also seems to be important during the development of tissues in the development of embryos and the healing of wounds.

10.6   Membranes

Cells are confined and defined by an outer membrane, whose major component is a lipid bilayer. Moreover, a eukaryotic cell is further subdivided into organelles, each of which also is confined within a membrane; in addition such cells also contain a labyrinthine network of membrane surface known as the **endoplasmic reticulum**. The importance of membrane-forming lipids in living cells is understated by their contribution in terms of simple mass; for example, they account for about 5% by weight of a mammalian cell, and 2% of a bacterium. But the total area of membrane in a liver cell with a volume of $5000 \, \mu m^3$ is $110000 \, \mu m^2$. Only 2% of this total area of membrane comes from the bounding membrane of the cell, the plasma membrane. This means that on average the separation of membranes is about 50 nm, not that much bigger than the size of a large macromolecule. Cells are systems that are dominated by surfaces, and these surfaces are provided by self-assembled lipid bilayers.

In Section 8.2.6, we saw that to form a stable bilayer, one needs amphiphiles with **large** hydrocarbon volumes and **small** values of the critical chain length and optimum head-group area. The major components of cell membranes—the phospholipids—fulfil this requirement. These molecules have two hydrocarbon chains, each with 12–20 carbon atoms linked through a glycerol phosphate unit to a polar head-group. One consequence of there being two hydrocarbon chains on each lipid is that their critical micelle concentration is extremely low; bilayers and vesicles formed from these molecules are highly stable. On the other hand, within one of the layers of the vesicles there is a substantial amount of mobility; in-plane diffusion coefficients are typically of the order 10^{-12} cm^2 s^{-1}. The relatively large size of the hydrophilic head-group also has important consequences; these head-groups are rather insoluble in the hydrophobic interior of the membrane. The probability of a lipid molecule migrating across the bilayer is therefore rather small; an upper bound for the rate has been estimated as 2×10^{-5} s^{-1}. As we shall see, these relatively small probabilities mean that the compositions of the two sides of a membrane may be substantially different.

Our first picture of a cell membrane is of a fluid sheet, stable with respect to the dissolution of its component lipids into the surrounding water, but dynamic in the sense that the component lipids have a high degree of mobility within the plane of the bilayer; each half of the bilayer is in effect a two-dimensional fluid. As we saw in Section 8.2.7, we should expect the membrane to have elastic properties which permit bending; at equilibrium we expect Brownian motion to lead to substantial thermal fluctuations in shape, which can be understood in terms of the bending energy formalism introduced in that section.

However, the picture of the membrane as a simple lipid bilayer needs some serious qualifications; as ever life has evolved structures of considerable complexity. A cartoon of the cell membrane of a eukaryotic cell is shown in Fig. 10.11. Some of the differences between a biological membrane and a synthetic vesicle that might be formed from a single bilayer-forming amphiphile are given below.

- Rather than being composed of a single amphiphile, biological membranes are made up of rather complex mixtures of phospholipids and other amphiphiles. Different types of membranes have different compositions, and there may be phase separation within the plane into regions of different composition.

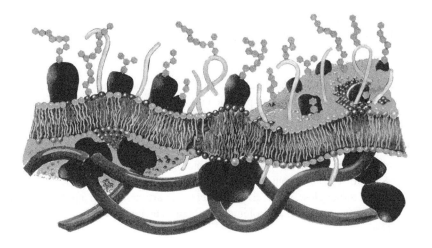

Fig. 10.11 A schematic diagram of the cell membrane of a eukaryotic cell. The basic structural element is a lipid bilayer; the bilayer is decorated by a number of protein molecules. On the inside of the membrane (below) a network of actin filaments is anchored by protein molecules, while the outside of the membrane is decorated by carbohydrate chains. Reproduced with permission from Mouritsen and Andersen (1998).

- In addition to the bilayer-forming amphiphiles such as the phospholipids, membranes often contain substantial amounts of other membrane-modifying molecules. In particular, around 25% of the membranes of animals can be composed of cholesterol, a molecule which by itself is only weakly amphiphilic, but which modifies the stiffness and permeability of the lipid membranes within which it is incorporated.

- Membranes also are associated with a number of different types of protein. Some of these **membrane proteins** are firmly anchored within the membrane; some completely cross it. Such **transmembrane proteins** may form a pore that allows molecules of the right size to cross the membrane; in some cases molecules may be pumped through the pores in opposition to a chemical potential gradient.

- Membranes are asymmetric, with a different composition inside than outside.

- The outside of the membrane is decorated by short polysaccharide chains. These may form the hydrophilic groups of certain amphiphiles—glycolipids—or they may be covalently bound to membrane proteins.

- The inside of the membrane may be linked to the cytoskeleton, the network of reversibly polymerised proteins such as actin. Links between the cytoskeleton and the membrane potentially allow cells to control the elasticity and curvature of the membrane, and underlie the ability of cells to move and divide by controlling the membrane's shape.

It is widely believed that at the beginnings of life vesicles were spontaneously formed from abiotically synthesised lipids. These vesicles served to compartmentalise the biochemical reactions of early life and formed the prototype of cells (see for example Maynard Smith and Szathmáry 1995). However this process happened, it is clear that membranes have now evolved to an immeasurably higher level of sophistication. Nonetheless the basic physical principles which underlie the behaviour of any self-assembled, flexible membrane must still apply.

Further reading

There are many excellent textbooks on molecular and cell biology; Alberts *et al.* (1994) is both comprehensive and readable. A brief, but highly attractive visual introduction to cell biology is provided by Goodsell (1998).

The classic textbook on the application of physical methods to biochemistry is Cantor and Shimmel (1980), which although beginning to show its age still contains much valuable material. Daune (1999) gives thorough and mathematical coverage to many of the properties of biological macromolecules. Grosberg and Khokhlov (1994) contains a number of applications of polymer theory to biological molecules. A collection of interesting articles giving a physics-based perspective on a number of biological problems can be found in Flyvbjerg *et al.* (1997).

Exercises

(10.1) a) A typical protein might have 100 amino acids linked together in a definite sequence. How many different sequences are possible for a polypeptide of this length made up of any combinations of the 20 common natural amino acids?

b) Suppose you have a method of synthesising 100 unit proteins which copolymerises the 20 amino acids completely at random. Estimate how much material you will have to synthesise in order to be sure of having at least one molecule of a given protein sequence.

(10.2) Calculate the critical micelle concentration for the phospholipid DPPC, which has two C_{16} chains attached to a phosphatidylcholine head-group. The energy required to transfer a lipid molecule from the bilayer to water is around 75 kJ mol^{-1}. Estimate how much larger the CMC would be if one of the chains was removed.

Some results from statistical mechanics

A.1 Entropy and the second law of thermodynamics

Statistical mechanics deals with systems made up of many interacting components. In such systems, one can have many different possible arrangements of the component parts. Each such arrangement is called a **microstate**. However, experimentally we are unable to measure the precise arrangement of all the components of the system (which in any case usually will be rapidly changing in time). Instead, we measure more global quantities which are essentially statistical in nature. For example, given a gas in a box we are unable to measure the positions and momenta of all the atoms of the gas, but we can measure the gas's pressure. To take another example from Chapter 5, if we have a solution of polymer molecules, we cannot measure the positions and momenta of all the polymer segments, but we can measure the average size of the polymer molecules. If the system has a well-defined value of such a global quantity, we say it is in a certain **macrostate**.

If we know something about the possible microstates of a system, can we predict its macrostate? Yes, because a given macrostate usually can be achieved by a number of different microstates. There are many different ways in which the segments of a polymer chain, for example, can be arranged in such a way that the chain has given overall dimensions. We call the number of different microstates consistent with a given macrostate the **statistical weight**, Ω. If all the microstates are equally likely, then the system is most likely to be found in the macrostate with the greatest statistical weight.

We define another quantity, the **entropy** S, in terms of the statistical weight:

$$S = k_B \ln \Omega. \tag{A.1}$$

An isolated system will, at equilibrium, be found to be in a state of **maximum entropy**; if the system is not at equilibrium any changes will cause the entropy to increase. This is a statement of the **second law of thermodynamics**.

Entropy can be thought of as a measure of **disorder**; there are more possible ways of arranging the components of a system when it is in a disordered macrostate than when it is in an ordered one. Suppose one keeps one's CD collection arranged in strict alphabetical order by artists; in this ordered state the collection has a statistical weight of one and a correspondingly low entropy. After the party, though, when the CDs have been removed and replaced at random, the collection is in a state with many possible equivalently disordered arrangements, and the entropy of the collection is much higher. If a system is not in a state of maximum entropy, it is not in complete **equilibrium**. The system will evolve until its entropy is a maximum possible value, at which time it will have reached equilibrium. This is a statement of the second law of thermodynamics.

Why is the entropy defined in terms of the log of the statistical weight? Suppose we have two independent subsystems, which for given macrostates have multiplicities Ω_1 and Ω_2, and corresponding entropies S_1 and S_2. Considering the two subsystems together as a single composite system, the statistical weight of the composite system is $\Omega_1 \times \Omega_2$, while its entropy is $S_1 + S_2$. Entropy is thus an additive quantity.

We can also write the entropy in terms of probabilities. We assumed above that all the microstates are equally likely, so the probability of finding the system in any given microstate $p = 1/\Omega$, and the entropy $S = -k_B \ln p$. If we define the microstates in such a way that they have different probabilities p_i, this equation can be generalised to

$$S = -k_B \sum_i p_i \ln p_i, \tag{A.2}$$

where the sum is over all possible microstates.

A.2   Energy, entropy, and temperature

If two bodies are in thermal equilibrium with each other, they have the same **temperature**. This follows from the requirement that the entropy is a maximum. Suppose that the energies are E_1 and E_2, and the entropies of the two bodies are $S_1(E_1)$ and $S_2(E_2)$, each being a function of the energy. The total energy of the system, $E = E_1 + E_2$, is constant. At equilibrium, the total entropy $S = S_1 + S_2$ must be a maximum. This implies that

$$\frac{dS}{dE_1} = \frac{dS_1}{dE_1} - \frac{dS_2}{dE_2} = 0. \tag{A.3}$$

We identify the quantity dS/dE with the inverse temperature:

$$\frac{dS}{dE} = \frac{1}{T}. \tag{A.4}$$

From the principle of maximum entropy one can deduce that two bodies at equilibrium with each other must have the same temperature, and if they are at different temperatures heat flows from the body at higher temperature to the body at lower temperature until the temperatures are equalised.

Now let us imagine that one of the bodies (the **reservoir**) is very much larger than the other one (the **system**); the two bodies are at equilibrium but heat can be transferred between them. What is the probability that the system is in a single microstate with an energy E_s? We know that the sum of the energies of the system, E_s, and the reservoir, E_r, is constant: $E_s + E_r = E$. The probability that the system is in a single microstate with an energy E_s is simply proportional to the number of microstates of the reservoir with energy $E - E_s$, that is $\Omega(E - E_s)$. If the reservoir is very large compared to the system, we can write

$$\ln \Omega(E - E_s) = \ln \Omega(E) - E_s \frac{d \ln \Omega}{dE}; \tag{A.5}$$

since $d \ln \Omega / dE = 1/(k_B T)$ this gives us the probability $p(E_s)$ that the system is in a state with energy E_s

$$p(E_s) \propto \exp\left(-\frac{E_s}{k_B T}\right). \tag{A.6}$$

This is the **Boltzmann distribution**.

A.3 Free energy and the Gibbs function

The second law of thermodynamics tells us that at equilibrium, the entropy of an isolated system is maximised. Very often, however, we consider systems that are not isolated. For example, in the previous section, we considered a system that was able to exchange energy with a reservoir kept at a constant temperature. We often wish to be able to predict the equilibrium state of an experimental system at constant temperature; to do this from this statement of the second law we would need to consider the entropy both of the system and of the reservoir. It turns out that there is a very convenient alternative formulation of the second law which allows us to apply it to a system that can exchange energy with its surroundings without explicitly considering their entropy. This involves the **free energy**[1] F, defined as

$$F = E - TS. \tag{A.7}$$

[1]Also called the Helmholtz free energy.

If a system has a **constant volume** and a constant number of particles, but is able to exchange energy with a reservoir at constant temperature, then for the total entropy of the system and the reservoir to be maximised, the free energy must be **minimised**.

Often we do experiments in which the system is constrained to be at **constant pressure** P, allowing the volume V to vary. Here, once again, we can formulate the second law in terms of a quantity that refers only to the system, and not to the reservoir. This quantity is the **Gibbs function**[2] G, defined as

[2]Also called the Gibbs free energy.

$$G = E - TS + PV. \tag{A.8}$$

So, for a system held at constant pressure and with a constant number of particles, which is able to exchange energy with a reservoir at constant temperature, then for the total entropy of the system and the reservoir to be maximised, the Gibbs function must be **minimised**.

To summarise, to find the equilibrium state of a system at a given temperature, we need to write down the free energy (for systems at constant volume) or the Gibbs function (for systems at constant pressure) as a function of the relevant variables, and then find the values of the variables which minimise the free energy or Gibbs function.

We should finally note a practical point. If we have a system whose volume is independent of pressure, then the free energy and the Gibbs function will be equivalent (to within a constant). Such systems are known as **incompressible**. Most liquid systems approach this limit quite closely, so many theories of soft matter start out with this approximation.

A.4 The chemical potential

In the last section we considered systems which were able to exchange energy with their surroundings. In many cases, we wish to consider systems that can exchange not only energy, but also matter, with their surroundings. For example, at the surface of an open beaker of liquid molecules are continuously leaving the liquid to join the vapour, and vice versa. How do we know whether a system and a reservoir are at equilibrium with respect to exchange of particles? We can use an argument which is an extension of the one by which we introduced temperature. Let us write down the differential change in the total entropy of two bodies $dS = dS_1 + dS_2$ when energy and particles can be exchanged between them. We have

$$dS = \frac{\partial S_1}{\partial E_1} dE_1 + \frac{\partial S_2}{\partial E_2} dE_2 + \sum_i \frac{\partial S_1}{\partial N_{i1}} dN_{i1} + \sum_i \frac{\partial S_2}{\partial N_{i2}} dN_{i2} \quad (A.9)$$

where we have summed over all chemical species i. Since both the total energy and the total numbers of each kind of particles are conserved we have

$$dS = \left(\frac{1}{T_1} - \frac{1}{T_2}\right) dE_1 + \sum_i \left(-\frac{\mu_{i1}}{T_1} + \frac{\mu_{i2}}{T_2}\right), \quad (A.10)$$

where we have introduced the **chemical potential** of the ith species, μ_i, defined by

$$\mu_i = -T \left(\frac{\partial S}{\partial N_i}\right)_{E,V}. \quad (A.11)$$

The chemical potential plays an analogous role with respect to particle exchange to that of temperature with respect to energy exchange. For two bodies to reach equilibrium with each other, energy flows between them until their temperatures are equalised, and particles flow until their chemical potentials are equalised.

Using the relations of thermodynamics, we can deduce a number of equivalent definitions of the chemical potential that are more immediately useful than eqn A.11, which applies for conditions of constant energy and volume. In particular, for conditions of constant temperature, we find

$$\mu_i = \left(\frac{\partial F}{\partial N_i}\right)_{T,V} = \left(\frac{\partial G}{\partial N_i}\right)_{T,P}. \quad (A.12)$$

Further reading

There are of course many undergraduate textbooks on thermodynamics and statistical mechanics. One intermediate-level text that is particularly well suited as a preparation for soft matter physics is Chandler (1987), which, in addition to a very clear recapitulation of the fundamentals, presents much useful introductory material about phase transitions and the liquid state.

The distribution function of an ideal random walk

In Chapter 5 we showed that for a freely jointed chain the mean squared end-to-end distance was $\langle \mathbf{R}^2 \rangle = Na^2$, where a is the step size, and we cited the result for the distribution of possible end-to-end distances

$$P(\mathbf{R}, N) = \left(\frac{2\pi Na^2}{3} \right)^{-3/2} \exp\left(-\frac{3\mathbf{R}^2}{2Na^2} \right). \tag{B.1}$$

These results are valid for any ideal random walk—that is to say, a walk in which the directions of successive steps are uncorrelated. Here we derive the result for the probability distribution in two ways. Our first method takes a simple example of a random walk and directly enumerates the statistical weight as a function of end-to-end distance. Our second method is more general, and shows the connection between random walks and diffusion processes, by showing that the probability distribution obeys a **diffusion equation**.

B.1 Direct enumeration of the statistical weight

As a simple model of a random walk, consider a walk in one dimension, in which each step of length a_x can be in either the forward or backward direction. If there are a total N steps, of which N_+ are forward and N_- are backward, then we can write the end-to-end distance R_x in terms of N_+ and N_- as

$$R_x = (N_+ - N_-)a_x. \tag{B.2}$$

To find the statistical weight Ω_x corresponding to a given value of R_x, we need to write down the number of possible walks with N_+ forward steps and N_- backward steps, where N_+ and N_- are related to R_x by eqn B.2. From simple combinatorics we have

$$\Omega_x = \frac{N!}{N_+!N_-!}. \tag{B.3}$$

If N is large, we can rewrite this using Stirling's approximation $\ln x! = x \ln x - x$; this gives us

$$\ln \Omega_x = N \ln N - N_+ \ln N_+ - (N - N_+) \ln (N - N_+) \tag{B.4}$$

where we have used $N_- = N - N_+$. We can rewrite this in terms of $f = N_+/N$ to find

$$\ln \Omega_x = -N[f \ln f + (1 - f) \ln (1 - f)]. \tag{B.5}$$

This function has a maximum at $f = 1/2$; the highest statistical weight attaches to the situation in which there are equal numbers of steps in either direction. Situations in which there are very many more steps in one direction than the other are much less likely, so we use a Taylor expansion to approximate this function for small deviations from $f = 1/2$. Thus we have

$$\ln \Omega_x \approx \ln \Omega_x(1/2) + \left(\frac{1}{2} - f\right)\left(\frac{d \ln \Omega_x}{df}\right)_{f=1/2}$$
$$+ \frac{1}{2}\left(\frac{1}{2} - f\right)^2 \left(\frac{d^2 \ln \Omega_x}{df^2}\right)_{f=1/2}. \tag{B.6}$$

At $f = 1/2$, the first derivative is zero, and the second derivative has the value $-4N$, so we have

$$\ln \Omega_x = N \ln 2 - 2N \left(\frac{1}{2} - f\right)^2. \tag{B.7}$$

Now, we can use eqn B.2 to rewrite this in terms of R_x to find

$$\Omega_x \propto \exp\left(-\frac{R_x^2}{2Na_x^2}\right). \tag{B.8}$$

Now, let us consider a three-dimensional random walk in which the x-component of each step is either $+a_x$ or $-a_x$, the y-component either $+a_y$ or $-a_y$, and the z-component either $+a_z$ or $-a_z$. For $a_x = a_y = a_z$ this gives a walk in which each step takes one of eight directions with a step size a, where $a^2 = a_x^2 + a_y^2 + a_z^2 = 3a_x^2$. The statistical weight in three dimensions is simply the product of the weights for each dimension:

$$\Omega = \Omega_x \Omega_y \Omega_z \propto \exp\left(-\frac{R_x^2}{2Na_x^2} - \frac{R_y^2}{2Na_y^2} - \frac{R_z^2}{2Na_z^2}\right); \tag{B.9}$$

since $R^2 = R_x^2 + R_y^2 + R_z^2$ and $a_x^2 = a_y^2 = a_z^2 = a^2/3$ this gives us

$$\Omega \propto \exp\left(-\frac{3R^2}{2Na^2}\right). \tag{B.10}$$

The probability distribution is simply proportional to the statistical weight, so after ensuring that the probability distribution is properly normalised we recover eqn B.1.

B.2 Random walks and the diffusion equation

Suppose we have a random walk in which each step can be represented by one of z vectors $\mathbf{b}_i$, each of length b. The probability distribution $P(\mathbf{R}, N)$ represents the probability that after N steps, the walk has reached position $\mathbf{R}$. But to get there, the last step must have been one of the vectors $\mathbf{b}_i$, and knowing this we can construct a relationship between the probability distribution after the Nth step and that after the $(N-1)$th step. This relationship is

$$P(\mathbf{R}, N) = \frac{1}{z} \sum_{i=1}^{z} P(\mathbf{R} - \mathbf{b}_i, N - 1). \tag{B.11}$$

Now if N is very large we can write $P(\mathbf{R} - \mathbf{b}_i, N - 1)$ in terms of $P(\mathbf{R}, N)$ using a Taylor expansion; in this way we will obtain a partial differential equation for $P(\mathbf{R}, N)$ in which we treat N as a continuous variable.

The Taylor expansion gives us

$$
\begin{aligned}
P(\mathbf{R} - \mathbf{b}_i, N - 1) = P(\mathbf{R}, N) &- \frac{\partial P}{\partial N} \\
&- \frac{\partial P}{\partial x} b_{ix} - \frac{\partial P}{\partial y} b_{iy} - \frac{\partial P}{\partial z} b_{iz} \\
&+ \frac{1}{2} \left(\frac{\partial^2 P}{\partial R_x^2} b_{ix}^2 + \frac{\partial^2 P}{\partial R_y^2} b_{iy}^2 + \frac{\partial^2 P}{\partial R_z^2} b_{iz}^2 \right) \\
&+ \frac{1}{2} \sum_{\alpha,\beta=x,y,z;\ \alpha\neq\beta} \frac{\partial^2 P}{\partial R_\alpha \partial R_\beta} b_{i\alpha} b_{i\beta};
\end{aligned}
\tag{B.12}
$$

we need to substitute this into eqn B.11. The resulting expression looks unwieldy, but we can simplify it considerably using some properties of the step vectors $\mathbf{b}_i$. The first spatial derivatives in the second line disappear when we sum over all the step vectors, because

$$
\sum_{i=1}^{z} b_{i,x} = \sum_{i=1}^{z} b_{i,y} = \sum_{i=1}^{z} b_{i,z} = 0
\tag{B.13}
$$

must hold if the walk is not to be biased in one particular space direction. Similarly the cross-terms on the third line also vanish, as

$$
\sum_{i=1}^{z} b_{i\alpha} b_{i\beta} = 0; \quad \alpha \neq \beta.
\tag{B.14}
$$

Finally, we have

$$
\frac{1}{z} \sum_{i=1}^{z} b_{ix}^2 = \frac{1}{z} \sum_{i=1}^{z} b_{iy}^2 = \frac{1}{z} \sum_{i=1}^{z} b_{iz}^2 = \frac{b^2}{3}.
\tag{B.15}
$$

Thus we find

$$
\frac{\partial P(\mathbf{R}, N)}{\partial N} = \frac{b^2}{6} \nabla^2 P.
\tag{B.16}
$$

The probability distribution function for a random walk obeys a **diffusion equation**. We need to solve it for an initial condition which is that after 0 steps, the walk must be at the origin $\mathbf{R} = 0$. The appropriate solution is

$$
P(\mathbf{R}, N) = \left(\frac{2\pi N a^2}{3} \right)^{-3/2} \exp \left(-\frac{3\mathbf{R}^2}{2Na^2} \right).
\tag{B.17}
$$

Answers to selected problems

C.1 Chapter 2

(2.1) $6.0\,N\,m$

(2.2) $597.0\,N\,m$

(2.3) Young modulus $Y = -nm(\epsilon/a^3)$

(2.4) We need to calculate the dimensionless ratio $(\rho E)/(Y m)$, where ρ is the density, E the bond energy, Y the Young modulus and m the relative atomic (or molecular) mass (care is needed with the units). We expect this ratio to be of order unity; in fact we find a variation from 0.08 to 17, but for pairs of materials with similar bonding the ratios are relatively close.

(2.5) A plot of $ln(\eta)$ versus $1/T$ yields an activation energy $\epsilon = 2.63 \times 10^{-20}$ J. This compares to the latent heat per molecule of 6.8×10^{-20} J.

(2.7) We use the condition that the experimental timescale $\tau_{exp} = \tau$, the relaxation time, at the glass transition temperature T_g. If the glass transition temperature at timescale τ_1 is T_{g1}, and is T_{g2} at timescale τ_2, then

$$\ln\left(\frac{\tau_1}{\tau_2}\right) = \frac{B}{T_{g1} - T_0} - \frac{B}{T_{g2} - T_0}.$$

For $T_{g2} - T_{g1}$ small compared to $T_{g1} - T_0$ this can be simplified to

$$\ln\left(\frac{\tau_1}{\tau_2}\right) = \frac{B(T_{g2} - T_{g1})}{(T_{g1} - T_0)^2}.$$

(2.8) The lower bound on the glass transition temperature, T_0, is found from the condition that the entropy in the glass and crystal are equal. This gives $T_0 = 129.5 K$.

C.2 Chapter 3

(3.1) a) 300 K, b) 0.25 and 0.75, c) 0.35 and 0.65.

(3.2) We need the solutions of

$$\ln\frac{\phi}{1 - \phi} + \chi(1 - 2\phi) = 0.$$

For large χ ϕ will be small and $1 - \phi$ close to 1, so in this limit we can write

$$\ln\frac{\phi}{1 - \phi} + \chi(1 - 2\phi) \approx \ln\phi + \chi = 0.$$

Thus $\phi \approx \exp(-\chi)$. Since χ is the energy change in units of $k_B T$ of taking one molecule from an environment of pure material A and putting in pure material B, this is simply a Boltzmann factor.

(3.3) Hexane, 1.3×10^{-5}, Octane, 8.3×10^{-7}, Decane, 5.4×10^{-8}.

(3.4) The most uncertain quantity in the estimation is the coordination number z. Taking a value of 20 for this gives $\gamma \approx 0.035\,\mathrm{J\,m^{-2}}$, which is the right order of magnitude.

(3.5) We expect $I(Q, t) \propto \exp(2R(Q)t)$. Plots of $\ln(I(Q, t)$ versus t are linear for small t, allowing us to extract $R(Q)$. Following eqn 3.22 we plot $R(Q)/Q^2$ versus Q^2 expecting a straight line. For real data (as this is) one often sees deviations from linearity at low Q, for reasons that are not always clear.

(3.6) Set $\exp(-\Delta G^*/k_B T) = (5 \times 10^{13} \times N)^{-1}$, where N is the number of atoms in a droplet. This gives $\Delta G^*/k_B T = 69.2$, from which we find the solid/liquid interfacial energy $\gamma_{sl} = 0.13\,\mathrm{J\,m^{-2}}$, and the critical radius $r^* = 1.3 \times 10^{-9}\,\mathrm{m}$.

C.3 Chapter 4

(4.1) Terminal velocities: sand grain, $6.5 \times 10^{-3}\,\mathrm{ms^{-1}}$, polymer particle, $2.7 \times 10^{-8}\,\mathrm{ms^{-1}}$, virus, $1.1 \times 10^{-10}\,\mathrm{ms^{-1}}$.

Diffusion coefficients: sand grain, $4.3 \times 10^{-15}\,\mathrm{m^2\,s^{-1}}$, polymer particle, $4.3 \times 10^{-13}\,\mathrm{m^2\,s^{-1}}$, virus, $4.3 \times 10^{-12}\,\mathrm{m^2\,s^{-1}}$.

Diffusion times: sand grain, $5.8 \times 10^5\,\mathrm{s}$, polymer particle, $0.58\,\mathrm{s}$, virus, $5.8 \times 10^{-4}\,\mathrm{s}$.

(4.3) The force as calculated by the Van der Waals approach equals the Casimir force at a separation of 120 nm. Above this distance, the Van der Waals approach overestimates the force due to the neglect of the finite speed of light (retardation effects). Below this distance, the Casimir approach overestimates the force due to a neglect of the finite depth of penetration of an electromagnetic wave into the gold.

(4.5) To achieve a Peclet number of unity, the velocity must be $2.7 \times 10^{-3}\,\mathrm{ms^{-1}}$. Thus at practical brushing rates we should see substantial shear thinning.

C.4 Chapter 5

(5.1) a) 67 nm; b) 168 nm.

(5.2) Remember that one needs to convert the force/unstressed area to the true stress, because in the strained sample the actual area is less than the unstressed area. Density of crosslinks is $9.1 \times 10^{25}\,\mathrm{m^3}$.

(5.4) The relative molecular mass between crosslinks is $18\,500\,\mathrm{g\,mol^{-1}}$. If the relative molecular mass were doubled, the plateau modulus would be unchanged, but if it were decreased by a factor of 5 the polymer would be unentangled and there would be no rubbery plateau.

(5.5) a) $1940\,\mathrm{g\,mol^{-1}}$ (care required with units), c) $5 \times 10^{-22}\,\mathrm{m^2\,s^{-1}}$.

C.5 Chapter 6

(6.1) a) 2500 s; b) 7500 s; c) 7704 s.

(6.2) The Flory-Stockmayer theory predicts that the gel fraction at degree of reaction $\Delta f/2$ is one half the value at Δf. Assuming a non-classical exponent of 0.41, the corresponding ratio is 0.753.

C.6 Chapter 7

(7.1) Without the cubic term, this model yields a second order transition. With the cubic term, the transition is first order. The value of the order parameter at the transition, $S_c = w/2u$.

(7.2) The critical voltage is 11 V. The switching voltage would need to be somewhat larger.

(7.3) $\Delta F_g = 5.6 k_B T$.

C.7 Chapter 8

(8.1) The crystal thickness l^* is found to follow

$$l^* = 5.687 + \frac{196.6}{\Delta T}\, nm,$$

where ΔT is the undercooling.

The melting temperature of a crystal formed at 400 K is 406.0 K.

(8.2) For a crystal in which the chains are fully extended, we need an undercooling of 19.9 K. If the chains are folded in half, the undercooling needs to be 94.0 K. This is unrealistic, however, as it assumes that the chain can reverse direction between two carbon atoms. In practise a finite length of chain must form part of the disordered fold surface.

C.8 Chapter 9

(9.1) a) Spherical micelles, with average radius 1.37 nm and average aggregation number 36.

b) Express the free energy change in the form of eqn 9.9, and then use eqn 9.12 to find

$$\langle |N - M|^2 \rangle = \frac{9}{2}\frac{k_B T}{\gamma a_0} M.$$

ii. 5.9.

(9.2) The energy of an end is $8.4\, k_B T$, and the most probable aggregation number is 703.

(9.3) a) 56, b) 3.5 nm, c) less than 0.2 times, d) $\chi < 5 \times 10^{-3}$, width $\approx$ 9 nm.

(9.4) a) 90 nm. For part (b), use the phase diagram in fig. 9.13.

C.9   Chapter 10

(10.1) There about 10^{130} possible sequences. To be sure of having one of each sequence, one needs to make about 10^{107} kg!

Bibliography

Adam, G. and Gibbs, J.H. (1965). *J. Chem. Phys.*, **43**, 139.

Alberst, B., Bray, D., Lewis, J., Raff, M., Roberts, K., and Watson, J.D. (1994). *Molecular Biology of the Cell* (3rd edn). Garland Publishing, New York.

Avnir, D. (Editor) (1989). *The Fractal Approach to Heterogeneous Chemistry*. J. Wiley, New York.

Ball, P. (1999). *The Self-made Tapestry* (2nd edn). Oxford University Press, Oxford.

Barham, P.J., et al. (1985). *J. Mater. Sci.*, **20**, 1625.

Bassett, D.C. (1981). *Principles of Polymer Morphology*. Cambridge University Press, Cambridge.

Binder, K. (1991). In *Phase Transformations in Materials*. Edited by P. Haasen. VCH, Weinheim.

Cantor, C.R. and Schimmel, P.R. (1980). *Biophysical Chemistry* (2nd edn, 1991). W.H. Freeman, San Francisco.

Chaikin, P.M. and Lubensky, T.C. (1995). *Principles of Condensed Matter Physics*. Cambridge University Press, Cambridge.

Chandler, D. (1987). *Introduction to Modern Statistical Mechanics*. Oxford University Press, New York.

Chandrasekhar, S. (1992). *Liquid Crystals* (2nd edn). Cambridge University Press, Cambridge.

Colby, R.H., et al. (1987). *Macromolecules*, **20**, 2226.

Cowie, J.M.G. (1998). *Polymers: Chemistry and Physics of Modern Materials* (2nd edn). Nelson Thornes, Cheltenham.

Daoud, M. and Williams, C.E. (Editors) (1999). *Soft Matter Physics*. Springer Verlag, Berlin.

Daune, M. (1999). *Molecular Biophysics*. Oxford University Press, Oxford.

Diamond, R., et al. (1975). *J. Mol. Biol.*, **82**, 371.

Dickinson, E. (1992). *An Introduction to Food Colloids*. Oxford University Press, Oxford.

Dinner, A.R., Sali, A., Smith, L.J., Dobson, C.M., and Karplus, M. (2000). *Trends Biochem. Sci.*, **25**, 331.

Doi, M. (1995). *Introduction to Polymer Physics*. Oxford University Press, Oxford.

Doi, M. and Edwards, S.F. (1988). *The Theory of Polymer Dynamics*. Oxford University Press, Oxford.

Donald, A.M. and Windle, A.H. (1992). *Liquid Crystalline Polymers*. Cambridge University Press, Cambridge.

Elliott, S.R. (1990). *Physics of Amorphous Solids* (2nd edn, 1991). Longman, Harlow.

Faber, T.E. (1995). *Fluid Dynamics for Physicists*. Cambridge University Press, Cambridge.

Flyvbjerg, H., Hertz, J., Jensen, M.H., Mouritsen, O.G., and Sneppen, K. (Editors) (1997). *Physics of Biological Systems*. Springer Verlag, Berlin.

de Gennes, P.G. (1979). *Scaling Concepts in Polymer Physics* (2nd edn, 1991). Cornell University Press, Ithaca, NY.

de Gennes, P.G. and Prost, J. (1993). *The Physics of Liquid Crystals* (2nd edn). Oxford University Press, Oxford.

Goodsell, D.S. (1998). *The Machinery of Life*. Springer Verlag, New York.

Grosberg, A.Yu. and Khokhlov, A.R. (1994). *Statistical Physics of Macromolecules* (2nd edn). American Institute of Physics, New York.

Grosberg, A.Yu. and Khokhlov, A.R. (1997). *Giant Molecules* (2nd edn). Academic Press, San Diego, CA.

Hamley, I. (2000). *Introduction to Soft Matter* (2nd edn). J. Wiley, Chichester.

He, C. and Donald, A.M. (1996). *Langmuir*, **12**, 6250.

Illett, S.M., Orrock, A., Poon, W.C.K., and Pusey, P.N. (1995). *Phys. Rev. E*, **51**, 1344.

Israelachvili, J. (1992). *Intermolecular and Surface Forces* (2nd edn). Academic Press, London.

Jäckle, J. (1996). *Rep. Prog. Phys.*, **49**, 171.

Jimenez, J.L., Guijarro, J.I., Orlova, E., Zurdo, J., Dobson, C.M., Sunde, M., and Saibil, H.R. (1999). *EMBO J.*, **18**, 815.

Jones, R.A.L. and Richards, R.W. (1999). *Polymers at Surfaces and Interfaces*. Cambridge University Press, Cambridge.

Krieger, I.M. (1972). *Adv. Colloid Interface Sci.*, **3**, 111.

de Kruif, C.G., van Iersel, E.M.F., Vrij, A., and Russel, W.B. (1986). *J. Chem. Phys.*, **83**, 4717.

Lifschitz, E.M. and Pitaevskii, L.P. (1981). *Physical Kinetics*. Butterworth–Heinemann, Oxford.

Lodge, T.P. (1999). *Phys. Rev. Lett.*, **83**, 3218.

Mahanty, J. and Ninham, B.W. (1976). *Dispersion Forces*. Academic Press, London.

Matsen, M.W. and Bates, F.S. (1996). *Macromolecules*, **29**, 1091.

Mau, S.C. and Huse, D.A. (1999). *Phys. Rev. E*, **59**, 4396.

Maynard Smith, J. and Szathmáry, E. (1995). *The Major Transitions in Evolution*. Oxford University Press, Oxford.

Mouritsen, O.G. and Andersen, O.S. (1998). *In Search of a New Biomembrane Model*. Biologiske Skrifter, Danske Videnskabernes Selskab, Copenhagen.

Mullins, W.W. and Sekerka, R.F. (1964). *J. Appl. Phys.*, **35**, 444.

Perkins, T.T., Smith, D.E., and Chu, S. (1997). *Science*, **276**, 2016.

Russel, W.B., Saville, D.A., and Schowalter, W.R. (1989). *Colloidal Dispersions*. Cambridge University Press, Cambridge.

Safran, S. (1994). *Statistical Thermodynamics of Surfaces, Interfaces and Membranes*. Perseus Books, Cambridge, MA.

Stauffer, D. and Aharony, A. (1994). *Introduction to Percolation Theory* (2nd edn). Taylor and Francis, London.

Strobl, G.R. (1997). *The Physics of Polymers* (2nd edn). Springer Verlag, Berlin.

Tabor, D. (1991). *Gases, Liquids and Solids* (3rd edn). Cambridge University Press, Cambridge.

Woodcock, L.V. (1997). *Nature*, **385**, 141.

Zarzycki, J. (1991). *Glasses and the Vitreous State*. Cambridge University Press, Cambridge.

Index